W9-CGY-376

FOUNDATIONS OF ASTRONOMY
From Big Bang to Black Holes

FOUNDATIONS OF ASTRONOMY

From Big Bang to Black Holes

Richard Knox

A HALSTED PRESS BOOK

JOHN WILEY & SONS
New York

TO MY MOTHER

Published in the U.S.A.
by Halsted Press, a Division of
John Wiley & Sons, Inc., New York.

Library of Congress Cataloging in Publication Data
Knox, Richard.
Foundations of astronomy.

"A Halsted Press book."
Bibliography: p.
Includes index.
1. Astronomy – Popular works. I. Title.
QB44.2.K58 1979 520 79-13
ISBN 0-470-26638-4

Printed in Great Britain

Contents

Standing on our microscopic fragment of a grain of sand, we attempt to discover the nature and purpose of the Universe which surrounds our home in space and time. Our first impression is something akin to terror. We find the Universe terrifying because of its vast meaningless distances, terrifying because of its inconceivably long vistas of time which dwarf human history to the twinkling of an eye, terrifying because of our extreme loneliness, and because of the material insignificance of our home in space—

THE MYSTERIOUS UNIVERSE (1930) Sir James Jeans (1877–1946)

Introduction: A Sense of Space

There is a well-known quiz question: on a clear day, how distant is the furthest object that you can see from the top of a 100m cliff?

The amateur surveyors among us will recall a rule of thumb which says that the horizon is at a distance $d = 3\frac{1}{2}\sqrt{h}$, where d is in kilometres and h is in metres, so that the answer is about 35km. The astronomers, however, will give the correct answer of 150 million kilometres. This is the distance of the Sun, which—as the question implies clearly enough—is visible from the top of the cliff.

To many people the Sun is an object to be worshipped, literally in some cases, but by many others it is worshipped figuratively in an annual ritual of sunbathing. It is not given another thought. When an occasional partial eclipse occurs, or even more spectacular (but alas very rare for most of us) a total eclipse, people marvel, but many still without the least idea that the Moon is passing between their line of sight and the Sun. The Moon, after all, appears to be about as big as the Sun, so it is probably about the same distance away, they may argue.

While the more informed amateur astronomer may be inclined to scoff at such naivety, he should bear in mind that subjects unfamiliar to a person are normally first learned through experience, unless the person deliberately sets out to learn about a subject in academic isolation. The first course is like learning a language by simply going to the country in which it is spoken and struggling along until the pressures of necessity and experience have produced some degree of practical familiarity with the language. The bookworm acquires a good idea of the grammar, the vocabulary and other such basic things, but is largely at a loss in conversation with a native.

For this reason, very detailed observations and amazingly accurate predictions of complex astronomical events were made by the ancient astronomers with what now seems incredibly slow progress towards understanding what was really happening. As will be seen in Chapter 1, this does not matter, as long as the theories held are not regarded as sacrosanct and immutable no matter what new suggestions or evidence may be put forward. The navigator can make his way by the stars with great precision, assuming them to be fixed on an infinite celestial sphere which rotates above the Earth once in just less than a day. Virtually no one would say that, as the evening draws near, the Earth is turning eastwards away from the setting Sun. We say merely that the Sun sets in the west.

But the tremendous obstacles to progress in understanding astronomy brought about by the almost religious adherence to the ideas of Ptolemy for over a millennium should have taught mankind a lesson: whatever model is adopted for a concept of the Universe, so long as the observed facts support it, it is worth developing—but one must never forget that it is a model.

So we can excuse the person who thinks that the eclipsed Sun is passing into the Moon's shadow, or that the stars shine by reflected sunlight, or that the Sun and Moon rise vertically over the horizon exactly in the east; and likewise we may forgive the poor observational powers of the artist who draws the slim crescent Moon with stars within the 'horns', or with the horns pointing to the lower right. People who have not asked 'Why?' will not remember what they have seen every day, but have never *observed*.

Even when dealing with the limits of the Universe, there is much that the observer can see for himself on the way to grasping the most difficult (and in most cases, still unconfirmed) ideas about its origins and contents. By relating as much as possible of what we learn to simple practical examples, the progress to understanding is very much easier.

It is the intention of this book to provide an aid equivalent to learning a language by visiting the country where it is spoken with a hand-book to the grammar and vocabulary provided to complete the theoretical needs. By spending a few minutes, but (most important these days) no money, the reader can link something he can see or do with much of the material covered. Even in the most difficult areas of cosmology, it is almost true to say that any individual's ideas on, for example, the origin of the Universe (which, strictly speaking, is cosmogony) are as good as any other's as long as they do not defy basic laws of science—and even these can change.

Astronomy is the oldest science, and yet such is its scope that it is also the one in which the least progress has been made. In fact it has been said, with some justification, that more progress has been made in astronomy in the last thirty years than in the previous three thousand. This is certainly true for planetary science, which has been changed completely with the advent of manned and automatic space vehicles. In the study of more distant objects, completely new techniques such as X- and gamma-ray astronomy, and indeed the study of the entire electromagnetic spectrum rather than merely the narrow band we see as light, have changed our basic knowledge completely. But for all that progress, astronomy is still at its dawning.

It is this which makes the subject so fascinating to the layman, and which is one of the reasons why there are so many highly skilled amateur astronomers. There is something for everyone to enjoy in the study of the Universe, whether or not he is a practising amateur astronomer. For most of us, the first good look at the Milky Way through even a modest pair of binoculars held on a rigid support is the time from which we are completely hooked.

The feeling of wonder grows with even a small telescope as you discover that not only are there thousands upon thousands of stars quite invisible to the naked eye, but also galaxies, nebulae and an amazing variety of objects. The beauty of the Universe is indescribable, but its impact never fades, and appreciation grows with understanding.

So that is what this book is about: it attempts to describe the present thoughts of astronomers about the nature and origins of the Universe itself; it describes the most important objects which have been found (and some which haven't!); it discusses the stars themselves and the galaxies in which they are grouped; it looks at the only good examples of planetary bodies that have been identified with any certainty or that have been studied at any length, those which circle the nearest star, the Sun which we see only 150 million kilometres away, on a clear day.

Richard Knox
Effingham, Surrey. 1978

1 The Outward Urge

The origins of astronomy are as old as the origins of Man. The earliest signs of civilisation—for example, the beginning of social practices such as hunting and (later) the cultivation of food crops—were linked to the seasons. There was a good time to sow seeds, and a plentiful time to hunt beasts. These seasons were clearly linked with the weather and the Sun, and quite naturally many men began to worship the Sun as the bringer of life.

It is easy to see how the Sun's pale twin, the Moon, then became deified. It was a vital god (or usually goddess) for the hunter, and played an important role at harvest time when the Moon near full would be in the sky night after night lighting the harvesters' return home.

So men, and in particular the priests, began to study the Sun and Moon, and the other celestial bodies. This observation continued with what now seems amazing persistence and astonishing accuracy for centuries, as the great civilisations grew and declined in the Middle East and later in Greece and Rome. By observations painstakingly recorded over hundreds of years the ancient Egyptian astronomers discovered a pattern to solar eclipses. From any one area of the Earth the size of Egypt, total eclipses of the Sun may typically be seen only once in eighty years. The data the Egyptian astronomers worked with had to be accumulated over a very long time indeed. Small wonder it was a task reserved for the select body of the priesthood: not only did the priest's prediction of an eclipse, and his rituals to ensure its passing, seem a miracle to the people, it *was* an amazing piece of practical astronomy. It was an art that was virtually lost for the best part of a thousand years after the decline of the ancient Egyptian civilisation.

Much of the knowledge thus gained by the Egyptians was used by the next torchbearers of astronomy, the Greeks. After them, astronomy came to a halt which lasted through the

Middle Ages until the Renaissance brought with it the fathers of modern astronomy.

But, before we look at the Universe as we now see it, we may ask, is it possible to recapture the view of the Universe held by the early civilisations? Except for the fact that most moderns know (that is, they have been told so, many times) that the Earth goes round the Sun, they are usually only aware that the Sun appears to go round the Earth each day. This is something that cannot be avoided; it is simply a matter of experience. If you put your chair in the sunniest spot in the garden and settle down to read this book, before long you will find that the Sun has moved round and maybe left you in the shadow of a tree. This is our common experience, and our senses tell us that the Sun is moving across the sky.

If we ignore the Sun and study the sky itself, we get the impression of a flat surface wherever we look. The sky clearly meets the horizon at right angles to the surface of the Earth, and yet crosses overhead and does the same at the horizon behind us. The impression is of a flattened dome over our heads across which the Sun travels.

This illusion is modified at night, when we see familiar patterns of stars together with the Moon clearly fixed on the sky and moving as if the whole dome were rotating. Moreover, the apparent increase in size of the Moon's disc, and the spread of the groups of stars (called *constellations*), when they are near the horizon gives the impression that the sky overhead is even further away than the horizon. It should be mentioned here that the apparent increase in size of the rising or setting Sun and Moon is entirely illusory, as can be checked quite simply by taking a photograph of the rising full Moon, and later of the Moon high in the sky, and comparing the sizes of the images on the film.

The camera is a useful accessory for naked-eye astronomy, as well as for telescopic work, and the list of suggested further reading on page 177 includes references for more details of this fascinating study. In general, however, an exposure of 30 seconds with an aperture of f2 and a film speed of 150 ASA will capture all the stars readily visible to the naked eye without objectionable blurring of their point images due to the rotation of the Earth. (A prolonged exposure, say 40 minutes, particularly in the region of the Pole Star (see page 124) will give a spectacular demonstration of the rotation of the Earth.) To photograph the full Moon when it is rising and to capture some of the foreground and background as well, try about 5–10 seconds at f2 with 100 ASA film.

So by observation alone, and without any preconceived ideas of the Sun's being at the centre of the Solar System, we reach the same obvious conclusions as the first observers of the sky; namely, that the sky with the stars fixed on it rotates about the Earth, and that each day the Sun crosses this vault of the heavens with his own independent motion.

The motion of the Sun is seen most convincingly by watching the shadow of the style on a sundial (or any shadow, if it comes to that—for example, the shadow which has crept across these pages while you were reading in the garden).

The measurement of time

Once the ancient astronomers started to measure the sky, it was soon clear that the Sun moved independently across the star sphere in the same way as the Moon, although the Moon made one circuit of the heavenly sphere every 28 days, approximately, and the Sun took one year.

This can be quickly demonstrated by setting up a simple telescope (see page 174) to point approximately south with a well-known bright star dead centre in the field of view and noting the time precisely. Leave the telescope fixed in position, and on the following day note the time that the same star is dead centre. This will then show the time the celestial sphere takes to revolve once about the Earth (in other words, the time for the Earth to make one revolution on its axis relative to the stars). This time is less than the time the Sun takes to make one revolution of the Earth by about 4 minutes.

The reason is simply that we regulate time by the Sun, for

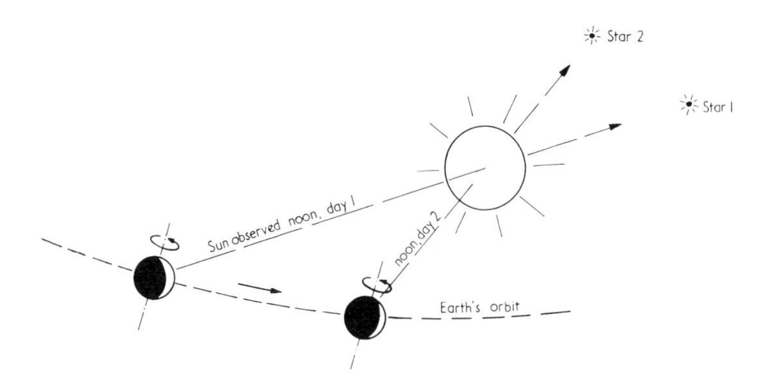

1 Owing to the movement of the Earth around its orbit, the stars behind the Sun at noon on one day are different from those on the next. The Sun apparently moves eastwards just less than 1° each day.

obvious reasons, and while the celestial sphere makes one revolution about the Earth (one *sidereal* day, i.e. star day) the Sun has moved eastwards against the background of the celestial sphere. Since it must complete a 360° revolution across the sky in one year, or 365 days, the Sun moves eastwards across the sky by just under 1° per day. This means that, relative to the stars, it will fall behind the stars by 1° each day, and it takes the Earth $24/360 = 0.067$ hours, that is 4 minutes, to cover this extra 1°.

So with the beginnings of measurement the Earth-centred Universe began to take shape. Elementary observation, carried out carefully with some simple angular measuring instrument such as the cross staff, soon showed that the Sun and Moon, and—more obviously—the planets, moved across the celestial sphere at an irregular rate. Sundials gain or lose during the year because the Sun gets ahead of or behind the position it would occupy if its motion were uniform. The Sun's error is small, some 17 minutes at worst, and so could be ignored by less time-conscious civilisations of the past, but when modern mechanical time-keeping became widespread the concept of the 'mean Sun', a fictional Sun with uniform motion across the sky each year, was introduced as the basis of 'mean solar time', or 'mean time' as we now call it.

The Moon was much more erratic: it could be 8° off-course, equivalent to over 30 minutes' error in the time calculated that it would be due south, for example.

The planets were even more confusing. Whereas the Sun and Moon merely speeded up or slowed down a little in their normal eastward motion across the sky, the planets would from time to time stop altogether, and even move backwards towards the west, making a loop or zig-zag in their course across the sky.

An explanation for such behaviour was important not only for academic and philosophical reasons, but also because, as astronomy became a more precise science in terms of measurement and positional work, so the growing traffic on the high seas began to use astronomy for navigation when out of sight of land.

It must be said, however, that the Greek astronomers cared little for such practical considerations as navigation, being occupied with philosophical matters such as the nature and size of the Universe. Thus this era marked the beginning of cosmology, the study of the Universe, and cosmogony, the study of the origins of the Universe. It was left to the Moorish merchants to apply the Greeks' knowledge of positional astronomy to navigation throughout the Middle Ages and, largely through this use, to preserve their knowledge for the Renaissance astronomers such as Copernicus, who was to mark the next major phase in the development of cosmology.

The Greek astronomers had to find an explanation for the erratic behaviour of the bodies other than the fixed stars. One famous early Greek astronomer, Aristarchus, of the third century BC suggested that the planets, including the Earth, went round the Sun. But his suggestion was soon forgotten with the work of Ptolemy, who represented the pinnacle of achievement of the Greek astronomers, and who formulated not only a coherent theory but backed it with measurements and predictions to show its validity.

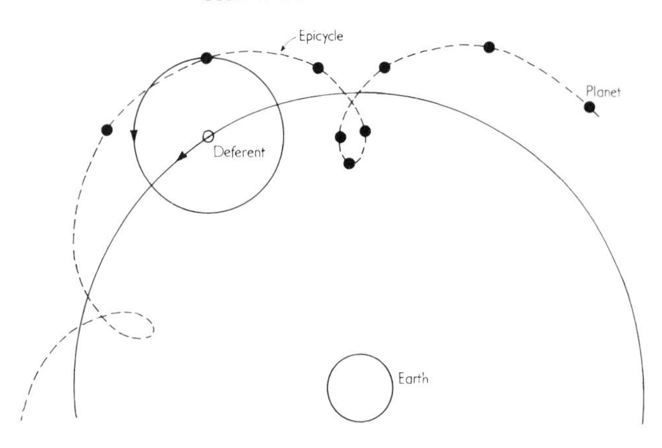

2 The complex mechanisms of deferent and epicycle needed by the Ptolemaic cosmology to explain the retrograde motion of planets across the celestial sphere.

Ptolemy, who lived between about 90 and 168AD, four hundred years after Aristarchus, said that the Universe must be ordered about circles, since these were held by the Greeks to be 'perfect' mathematical forms, and hence must be the forms used by the gods. The planets were therefore supported on crystalline spheres rotating about the Earth; but, to account for the non-uniform motion, these spheres were centred half-way between the Earth and an 'equant point', a purely fictitious point about which the planet had uniform angular motion. This was still not enough: the retrograde (that is, westward) motion of the planet across the sky could only be explained by making it rotate about another point, called a 'deferent', which was in orbit about the equant and Earth, the planet turning on a smaller orbit about the deferent, thus forming an epicycle. This was still insufficient to explain the observed positions in some cases, so further epicycles were added as necessary.

It may seem incredible now that such a complex system of 'wheels within wheels' could be accepted as true. But it must be remembered that it was regarded as less important to provide a true physical explanation than to find a way to predict the behaviour of the Sun, Moon and planets. This the Ptolemaic model did admirably. The positions of the planets can still be worked out with very creditable accuracy using the Ptolemaic system and no very complicated arithmetic.

The stars were less important in the Ptolemaic Universe since they were merely the unchanging pattern on the outer sphere which formed the impenetrable background to the crystal spheres of the planets.

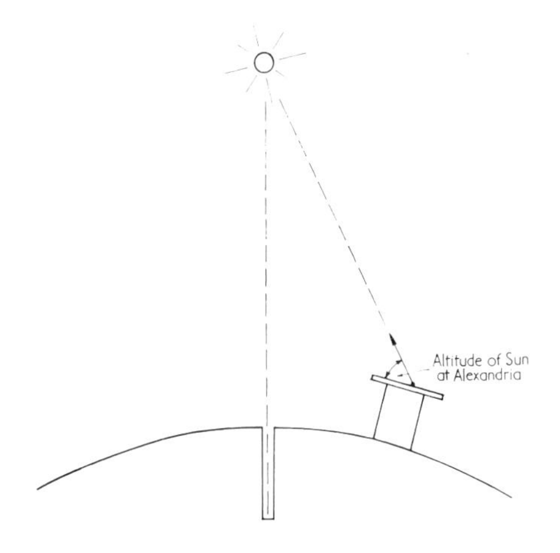

3 Eratosthenes' method for finding the radius of the Earth. The Sun overhead at Syene shone directly down a deep well, yet in Alexandria it was measured at 7.2° from overhead. From this difference, and knowing the distance between Alexandria and Syene, the radius of the Earth could be calculated.

Two other major achievements of the Greeks are worth mentioning, both from their practical importance and their relevance to cosmology. Eratosthenes, who lived at about the same time as Aristarchus, noted that the Sun reached maximum altitude at noon on the summer solstice exactly overhead at Syene in the south of Egypt, while the maximum altitude at Alexandria was 7.2° from the zenith. Now, since the distance between the two places was known with some accuracy, the radius of the Earth could be calculated by elementary geometry. There is some confusion over the exact answer which

Eratosthenes reached as we are uncertain of the exact value of the *stadium*, the distance unit used. It appears likely that his estimate was too small, but by not more than 7 per cent. The world had been sized up pretty closely!

Aristarchus devised an ingenious method of finding how far the Sun was from the Earth relative to the Moon. He measured the angular separation across the sky of the Sun and Moon when the Moon was exactly half illuminated (at the first quarter). This he made 87°, which meant, by simple triangulation, that the Sun was 19.1 times as far away as the Moon. This was a considerable underestimate, owing largely to the difficulty of determining exactly when the Moon is at first quarter, and to the precision required for the angular measurement. To arrive at the correct answer, the angle should have been found to be about 89.85°; not much different from a right angle.

However, Aristarchus had shown that the Universe was bigger than was thought: it has been growing ever since, both literally and in Man's estimation.

After the Greeks, astronomy and cosmology made no progress—in fact in the Europe of the Middle Ages, many backward steps were taken. The work of the Greeks was taken up and expanded by the Arabs, who left their mark in the names and terminology of modern astronomy, but was returned to Europe through other Middle-Eastern communities by the Moorish invasions of Spain and of other parts of Europe.

The Sun at the centre

With the Renaissance, European astronomy began to flourish once more. The most significant development was the overthrow of the Earth as the centre of creation. The Polish canon Copernicus almost diffidently suggested that if the Sun were taken as the centre of the Universe, the explanation of the irregular motion of the planets became much simpler than the cumbersome Ptolemaic machinery, since now the observer could interpret his measurements as if the Earth were also in orbit round the Sun.

This concept involved a major psychological upset, and, while treated with respect at the time, quickly fell into disrepute again. The problem was that the new Copernican model gave mathematical results less accurate than Ptolemy's, unless epicycles were added on to the orbits round the Sun, making the system just as cumbersome as Ptolemy's. Moreover, it offered no explanation of why the planets should be held to the Sun at all, so Copernicus retained the idea of crystalline spheres.

Copernicus' ideas were published at first privately in an abbreviated form in about 1510; then, after much persuasion by his disciple Rheticus and by Pope Clement VII, they were published in full in *De revolutionibus orbium cœlestium* (1543), as Copernicus himself was dying.

Three years later, Tycho Brahe, the greatest of the naked-eye astronomical observers, was born. In later life Tycho took a sceptical view of the Copernican system; but, as it offered certain advantages, he proposed his own version in which the Moon and Sun orbited the Earth (since these bodies display no retrograde motion) and the other planets orbited the Sun! Mercury and Venus both circled the Sun within the orbit of the Sun around the Earth, and hence were always seen close to the Sun in the sky, while the planets Mars, Jupiter and Saturn (which were all that had been identified at that time) had orbits of greater diameter than that of the Sun, so that they could pass the Earth on the opposite side to the Sun. Any cosmology of the Solar System had to allow the outer planets to reach *opposition* every year. This is when they are due south at midnight, exactly opposite the Sun in our sky.

But Tycho's greatest gift to progress in understanding the Universe was the precision of his observations. Tycho's measurements passed into the hands of his assistant, Johannes Kepler, upon Tycho's death. Kepler, working with Tycho's observations of the positions of Mars, which were particularly troublesome to predict by either Ptolemaic or Copernican schemes, made two fundamental changes to the models of all his predecessors: he saw that the 'perfect' circle would have to

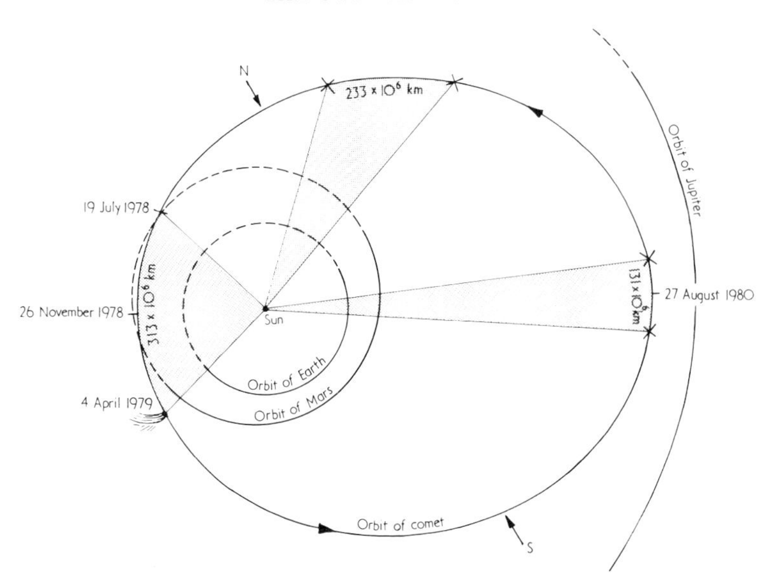

4 Orbit of comet Clark (1973 V) compared with the orbits of Earth, Mars and Jupiter. The diagram is drawn in the plane of the comet's orbit, and the planets' orbits are shown dotted when south of this plane. The comet's orbit is inclined at 9° to the plane of the Solar System (ecliptic) and it reaches furthest north and south at N and S respectively. The shaded areas illustrate Kepler's second law, being equal areas for periods of 260 days. The comet takes $5\frac{1}{2}$ years to complete one orbit. The eccentricity of the ellipse is about 0.5.

be abandoned in favour of an ellipse, and he proposed that the planets moved round their elliptical orbits at varying speeds.

This was almost the final abandonment of the Greek philosophy of the 'perfect' geometrical Universe. The only remnant to which Kepler clung was the heavenly spheres on which the planets were held in their orbits. At first he had hoped to show that these spheres were of just the right size to accommodate the regular solid figures of geometry, allowing for the thickness of the Copernican epicycles, but towards the end of his life he tried to show that they were in the same ratios as the notes on the musical scales that Pythagoras had believed made the 'music of the spheres'.

Kepler's passing marked the virtual separation of astronomy from astrology and other mystical associations. His three laws of planetary motion were particularly important because they apply to bodies in orbit around each other anywhere in the Universe, and hence enable astronomers to deduce a great deal about the physical properties of the stars. Yet these laws were formulated empirically from naked-eye observations made by another man: a great tribute to both Tycho and Kepler.

Kepler's laws are:

1 A planet orbits the Sun in an elliptical orbit, with the Sun at one focus of the ellipse.
2 The line joining the Sun to the planet (called the *radius vector*) sweeps through an equal area in any given period of time.
3 The square of the period of rotation about the Sun (the *sidereal period*) is proportional to the cube of the mean distance from the Sun.

The significance of Law 1 was the breaking of the perfect Greek circle. Law 2 meant that a planet close to the Sun would have to travel faster than when at a greater distance to sweep through an equal area. These first two laws are reminiscent of Ptolemy's off-centred deferent with the prime body at one focus and the empty equant point at the other, so perhaps we should not regard Ptolemy's system as so fantastic.

While Kepler was producing his own set of nails for the coffin of Greek astronomy, many more were being forged in Italy by Galileo Galilei, a mathematician of the University of Padua.

In 1608 a Dutch lens grinder, Johannes Lippershay, had invented spectacles and the telescope, the telescope being a combination of two spectacle-type lenses, one of long and the other of fairly short focal length. The following year, Galileo turned this new invention on the sky, and the most important step in observational astronomy had been taken. Immediately,

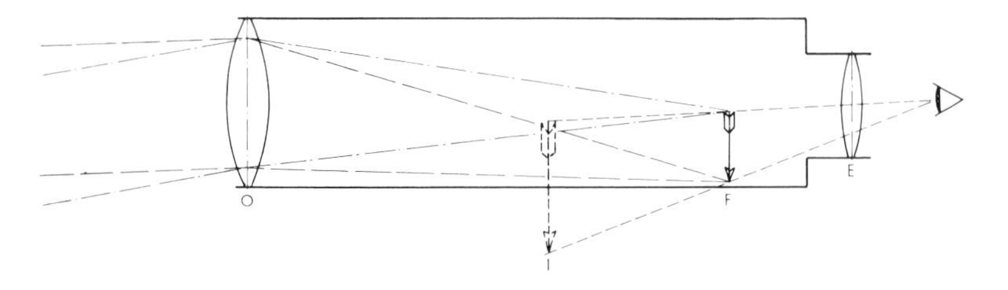

5 Principle of the simple refracting telescope. The objective, O, forms an image at F which when examined through eyepiece E produces an enlarged inverted image of the object being studied.

Galileo could see that Venus had phases like the Moon, which could only be caused by revolution about the Sun.

It was also immediately evident that the Earth need not be the centre of all orbits: Galileo saw four bright satellites clearly in orbit around Jupiter. This was the final proof of the Copernican cosmology. The story of how Galileo upset the Church, and how his critics refused even to look through Galileo's newfangled device, is a familiar tale of the fight between the pioneer in science and the establishment scientists. Such events have been repeated through the ages, and in all major branches of science. We need not concern ourselves here with this fascinating story, except to note that Galileo had initiated the biggest explosion in astronomical knowledge since the beginning of civilisation, a step unequalled until the invention of radio astronomy and space vehicles 350 years later.

The last key that was to free the intellectual bonds of Man's imagination, and to stimulate the urge to probe ever deeper into the Universe that has motivated astronomers and cosmologists ever since, was supplied by Sir Isaac Newton.

Copernicus had provided the basic theory, Kepler had refined it on the evidence of Tycho Brahe, and Galileo had provided the visual proof. Newton with his laws of universal gravitation, was to provide an unprecedented advance in the laws of physics and celestial mechanics, explaining Kepler's empirical laws, disposing of musical and geometrical explanations, and devices such as crystalline spheres, and laying the ground from which Neil Armstrong was to step on to the Moon.

Newton said, basically, that every particle of matter attracted every other with a weak force that varied as the square of the distance separating the particles. Added to this was his law that every particle continues in a uniform state of motion unless a force is applied to change that state (the force of gravity being one such). These laws of motion, verifiable by experiment, opened the gates of understanding.

The Moon and the Earth attract each other, though each would continue to move in a uniform manner without the other. Because of the greater mass of the Earth, this mutual force disturbs the Moon more than the Earth. An orbiting body constantly tends to fly off at a tangent from its orbit, as the Greeks knew, but the gravitational attraction of the primary body continually pulls the planet to the Sun, or the satellite towards the planet. The net result of this tug of war is that a satellite and planet revolve around the common centre of gravity of the system. The satellite speeds up as it approaches the primary, just as Kepler observed, so as to balance the gravitational force which is increasing as the inverse square of the distance. Similarly, a falling body (which is what an orbiting satellite is) will accelerate towards the centre of mass of the body attracting it due to the gravitational force attracting it. This is why the orbits of celestial bodies are ellipses, as Kepler also observed.

Such was Newton's contribution to our understanding of the Universe that it was not until Albert Einstein in the twentieth century that the next fundamental changes occurred. Many great astronomers meanwhile worked and discovered and invented. Ideas about the size, shape and origins of the Universe proliferated. The telescope showed that the stars were not merely points on some distant sphere, but varied in distance enormously. The stars, it was seen, were grouped in galaxies, many bigger than our own (not even the distinction of inhabiting the largest galaxy or even the centre of a galaxy was to be granted to Man), and, with the growth of radio astronomy and space-vehicle experiments, a new age of astronomy began. Numerous entirely new astronomical objects have been discovered, so that it seems that the truth of the Universe is stranger than the wildest dreams of the speculators.

To try to understand the Universe seems almost a blasphemy, as it did to the clerics who refused to look through Galileo's telescope, and as it did to Ptolemy who, content with what he could see, was happy with his simple answers. How surprised would Ptolemy, awakened today, be at the concept of the Universe proposed by Einstein? A Universe in which nothing is absolute: neither time nor any dimension, but only the speed of light; a Universe in which gravity is not a force but a distortion of space itself. Is such a Universe more strange, after all, than one in which the planets of the Solar System orbit the Sun in ellipses?

2 The Lonely Wanderers

The origins of the planetary system which we inhabit are linked, inevitably, with the origins of the Universe itself. But the history of astronomy and cosmology has shown time and time again that we have consistently overestimated the amount we know, and we have consistently underestimated the complexity, size and age of the Universe, and we have always been taken aback by the many complete surprises that have emerged. So it helps, in our quest to try to put the Universe into perspective a little more clearly, if we start at home and work outwards; in effect, retracing the methods and progress of astronomers through history.

Briefly, we can reassure ourselves that the Earth is a sphere, or near enough, by logic (the Moon and Sun are clearly spherical, and through a telescope we see the planets are too, so therefore the Earth is likely to be also), by observation (during a lunar eclipse, the Moon passes into the Earth's shadow, whose circularity clearly shows that the Earth is round) and by reliable evidence (from men in space).

We also saw that from the top of a cliff the horizon is further away than if we are on the beach: the rule of thumb is that the distance to the horizon in kilometres is $3\frac{1}{2}$ times the square root of the height of your eyes above sea level in metres. If you are 2m tall, for example, the horizon when you are paddling at the water's edge is about 5km distant. This rule of thumb is deduced from elementary geometry. It can be verified by simple tests (how far up a cliff you need to climb to see a known landmark across a stretch of sea) and depends for its proof on the fact that the Earth is more or less spherical.

Eratosthenes' method (see page 17) for measuring the diameter of the Earth is rather difficult for most people to repeat, owing to the need to make measurements of the Sun's altitude at different parts of the Earth's surface far enough apart. There is, however, a rough and ready way of checking for yourself. If

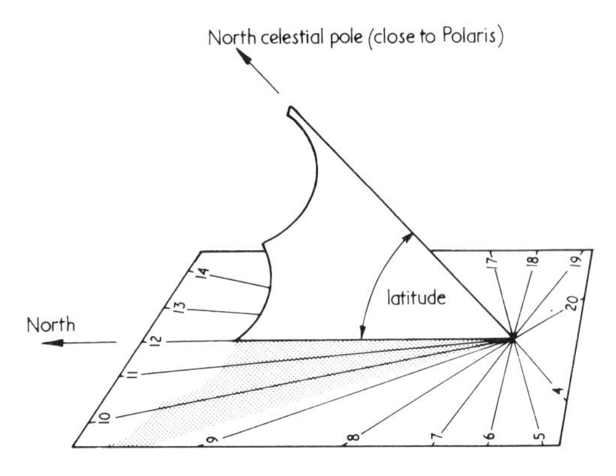

6 How an ordinary garden sundial indicates the north, the pole star, and the latitude of the place.

you go well to the south or north of your home, on a holiday for example, slip a protractor in your luggage. Then find a sundial at the place you are visiting. The *style* (the triangular shadow-casting device on the traditional garden sundial) must point at the celestial pole. This is the point in the sky towards which the Earth's axis is directed, and around which all celestial objects, including the Sun, move once each day owing to the Earth's rotation. The style points at the celestial pole so that, whatever the position of the dial on the Earth or the inclination of the dial plate to the ground, it is parallel to the Earth's axis, so that the Sun rotates uniformly about the style's shadow-casting edge.

All you have to do is measure the inclination of the edge of the style to the horizontal. Make a similar measurement on a dial near your home and you have two measurements of the altitude of the celestial pole at different latitudes. Since the lines through the places where we made the measurements and the celestial pole must be parallel, we can find the angle between the two places at the centre of the Earth by simple geometry. Now we measure the north–south distance between the places (i.e. the distance between their circles of latitude) from a map. The proportion of that distance to the circumference of the Earth is the same as the angle at the Earth's centre is to 360°. So we know:

$$\frac{d}{\pi D} = \frac{\Delta A}{360}$$

where d is the north–south distance between the sundials, ΔA the difference in the angles of the styles, and D is the Earth's diameter.

$$\text{Thus } D = \frac{d}{\pi} \times \frac{360}{\Delta A}$$

For example, in London, the inclination of a style is measured as 51½°, and in Valencia a sundial is found with the style at 39½° to the horizontal. The distance between them (north–south only) is 1340km, so the diameter of the Earth estimated from these figures would be 12 800km. (The equatorial diameter of the Earth is, in fact, 12 756km.)

So much for the Earth, the only planet we know for sure to be a home of intelligent life. Is it unique in the Universe in this respect? Many have argued plausibly that it is. But to suggest that Man is unique is the same in principle as suggesting that the Earth is the centre of the Universe. The Earth has a unique position in the system of planets it occupies, at least, in that it looks increasingly as though there is no life elsewhere in our Solar System, and the best we may hope for now is that some very rudimentary form of life may yet be found on Mars, or even, it has been suggested, in the atmosphere of Jupiter. We shall return to this subject later.

Another unique feature of the Earth in the Solar System is that it could be called a 'double' planet. The second body with which it shares its orbit, the Moon, is far bigger relative to the primary body than any other satellite in the Solar System. If we divide the diameter of the planet by the diameter of its largest satellite we get 3.7 for the Earth–Moon, compared with 295 for Mars–Phobos, 28.4 for Jupiter–Ganymede, 20.5 for Saturn–Titan, 26 for Uranus–Titania, and between 12 and 8 (depending upon the diameter of the satellite, which has been estimated at between 4000 and 6000km) for Neptune–Triton. Even though some uncertainty still remains over the sizes of the satellites of the outermost planets, it can be seen that the Earth, less than four times as big as the Moon, has a planet-sized companion.

The distance to the Moon

How did we know the distance to the Moon before the Apollo astronauts put down laser reflectors on its surface? The laser experiment, by timing the passage of the laser beam to and from the Moon, fixes its position to within a metre or so.

Having determined the diameter of the Earth, we know we have two positions on Earth over 12 700km apart from which to observe the Moon. Early astronomers found that the Moon's position, relative to the background of stars, could differ according to the observer's position on Earth. This is simply observed when the Moon passes in front of a particular star as seen from one place, but not from another. The early astronomers concluded, rightly, that this was due to parallax. The Moon is very much closer than the fixed stars, and is therefore apparently displaced in the sky according to the observer's geographical location.

At the extremes, from diametrically opposite points of the Earth, the Moon is displaced by some 2° (four times its apparent diameter) relative to the stars. So the Moon's distance of 384 400km can easily be deduced by parallax measurement. From a measure of its angular size, about 0.5°, you can now work out the diameter of the Moon. This is 3 476km.

So the Earth and Moon are globes nearly 400 000km apart, with diameters of about 13 000 and 3 500km, to use round numbers. If you imagine these to scale, they seem to be rather distant neighbours. A journey to the Moon is almost ten times the distance round the Earth.

The planets, as the ancient astronomers realised, are even more remote. Their apparently slow progress across the sky suggested enormous distances, and when measurement tech-

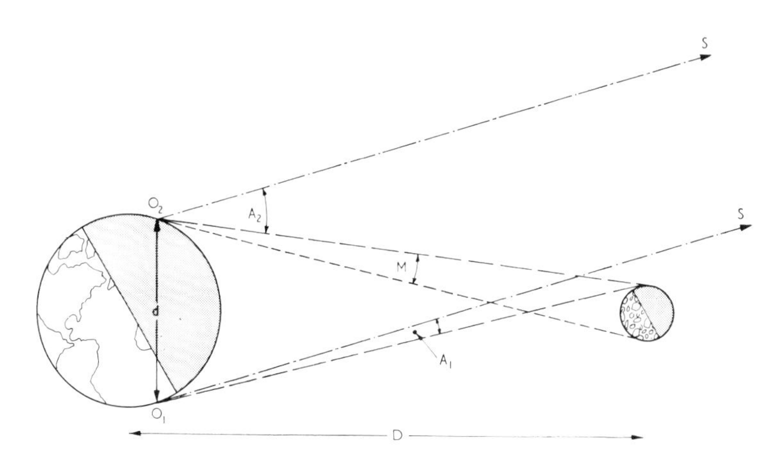

7 The Earth and Moon to scale but with their separation reduced by a factor of ten to exaggerate parallax (also approximately tenfold). Simultaneous observations from widely separated points on Earth (0_1 and 0_2) measure the angle between one side of the Moon and the star S. The difference (A_2–A_1) and the known separation of the observers (d) allows the distance to the Moon (D) to be calculated. The angle M then gives the Moon's diameter.

niques had been perfected, this was found to be true. The numbers we begin to encounter involve many thousands, millions, or thousands of millions.

We can conveniently indicate such big numbers by using powers of ten to multiply more manageable numbers. For example, the distance to the Sun can be expressed as 150×10^6 km, which means 150 followed by six zeros (the actual figure is 149 600 000km). But even better, if we express all the interplanetary distances in terms of the mean Earth–Sun distance we find the numbers more manageable, and we have a direct comparison. So we call 149.6×10^6km one *Astronomical Unit* (1 AU).

How was the Earth–Sun distance (i.e. the Astronomical Unit) determined? This proved to be harder than was thought by early astronomers, because of the unexpectedly high value of the distance. Aristarchus, as we saw earlier, thought the Sun

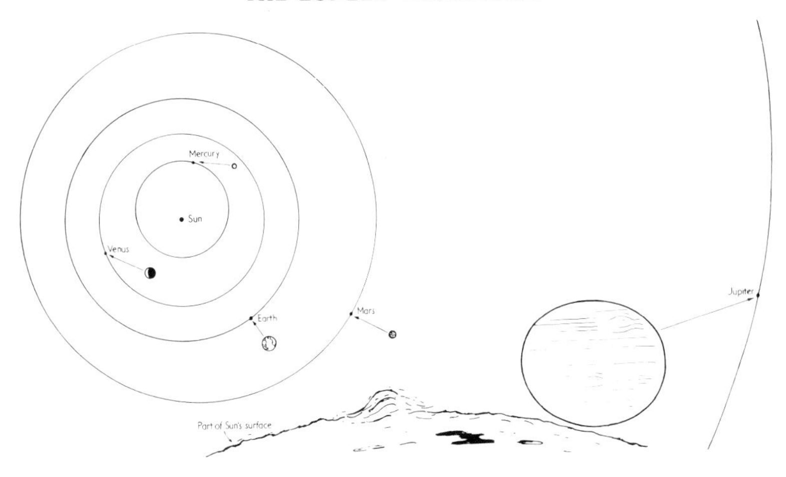

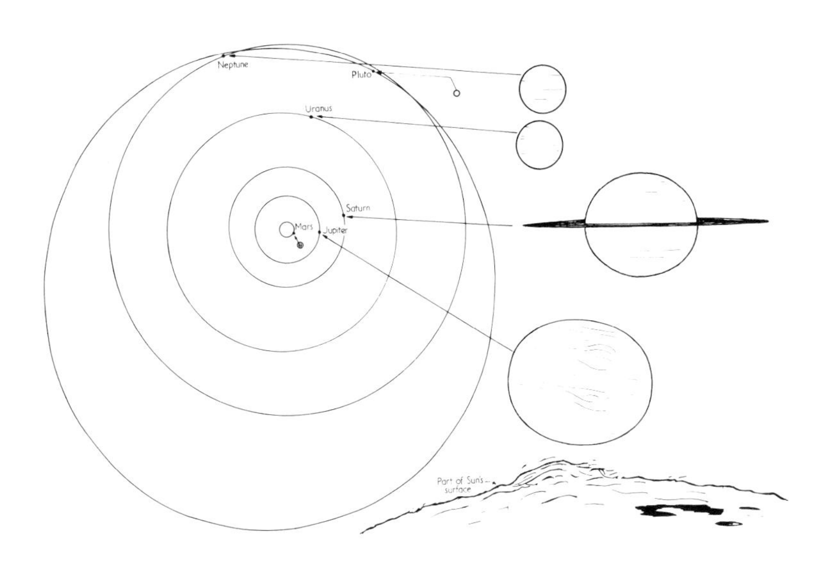

8 and **9** The planets shown in their relative positions on Jan 1, 1980, with the orbits to scale, and the planetary discs to a larger scale. The size of the Solar System is so vast that the outer planets' orbits are shown to a different scale in the second diagram. (The orbit of Mars is included in both for comparison.)

was 19 times the Moon's distance, which would have made the Astronomical Unit 7.3×10^6km, a 20-fold error.

The evaluation of the Astronomical Unit assumed an even greater importance after Kepler formulated his three famous laws, because the significance of the third law is that it gives a method of determination of all planetary distances once one has been fixed.

The square of the period of rotation, t, is *proportional* to the cube of the mean distance. Thus, for the Earth if we call the distance 1 astronomical unit, and the period of rotation 1 year, we have for the Earth $t^2 = d^3$. This will allow us to calculate the mean distance of any planet once we know its sidereal period, its period of rotation about the Sun, which we can easily determine by observation.

For example, what is the mean distance to Jupiter? Jupiter's sidereal period is 11.8 years. So $d = \sqrt[3]{11.8^2} = 5.2$ AU. What about Pluto, which takes 247 years? Its distance is an incredible 39.4 AU, and that is only the mean distance. It can be as much as 49 AU, owing to the eccentricity of the orbit. Pluto at its closest to the Sun, as it will be in 1989 for the first time since its discovery, is less than the mean distance of Neptune, and is only 29.6 AU from the Sun. The closest the Earth can be to Pluto is therefore 28.6 AU (4,279 million kilometres). The Solar System is 'full' of lonely planets.

As astronomical techniques improved, so the value of the astronomical unit was refined from Aristarchus' gross underestimate. Direct measurement of parallax of planets such as Mars, or one of the nearer asteroids when they began to be found, helped refine the Astronomical Unit, since, by application of Kepler's third law, once the distance was determined for any planet, all the others followed. Thus Kepler himself estimated the distance of the Sun at 22.5 million kilometres. Experiments were carried out by parallax measurements and by timing transits of the planet Venus across the disc of the Sun, comparing timings made 'simultaneously' on the other side of the world. Unfortunately, although this could give a reasonable value in theory when the method was just devised, once more being a simple triangulation from the Earth, it depended upon precise co-ordination and extreme accuracy of timing, well beyond the capability of the best timepieces of the period.

Now we know the distance to the planets, and hence the Sun, by reflecting radar signals from the nearest. Thus the yardstick by which the distances to the stars are found has been forged.

Do-it-yourself solar system measurement

The yardstick can be determined by other means, some within the scope of the amateur once armed with one piece of the puzzle. For example, if you know that the speed of light, determined by non-astronomical methods, is 299 792.5km/s (300 000km/s if rounded up) you can fix the value of the AU by observing the satellites of Jupiter.

From Kepler's third law Jupiter's mean distance is 5.2 AU, and, by drawing the Earth's orbit and Jupiter's to scale in AU, you can simply determine the distance of Jupiter from Earth at any date (see diagram). Now on August 10 in 1978, you observe an eclipse of one of Jupiter's satellites, Europa, by means of a small telescope (see page 174) at 21^h 08^m. On December 23 of that year, you observe it enter eclipse at 22^h 08^m. On December 31, the eclipse begins at 00^h 42^m.

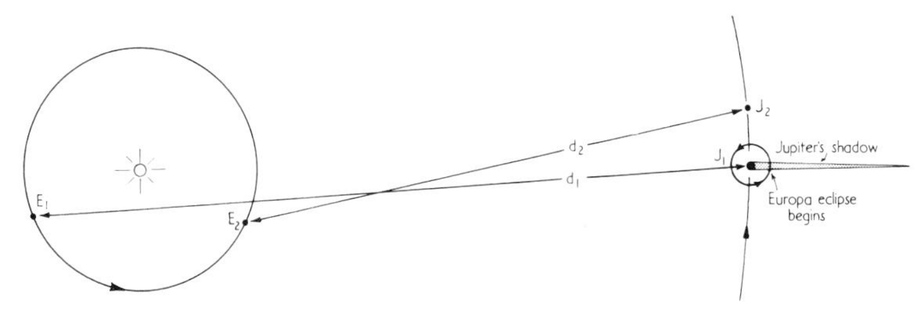

10 Owing to the difference in the distance from Earth to Jupiter (d_1–d_2) in August and December 1978, Europa's eclipses in Jupiter's shadow appear to occur slightly earlier than expected because of the time light takes to cover the distance d_1–d_2.

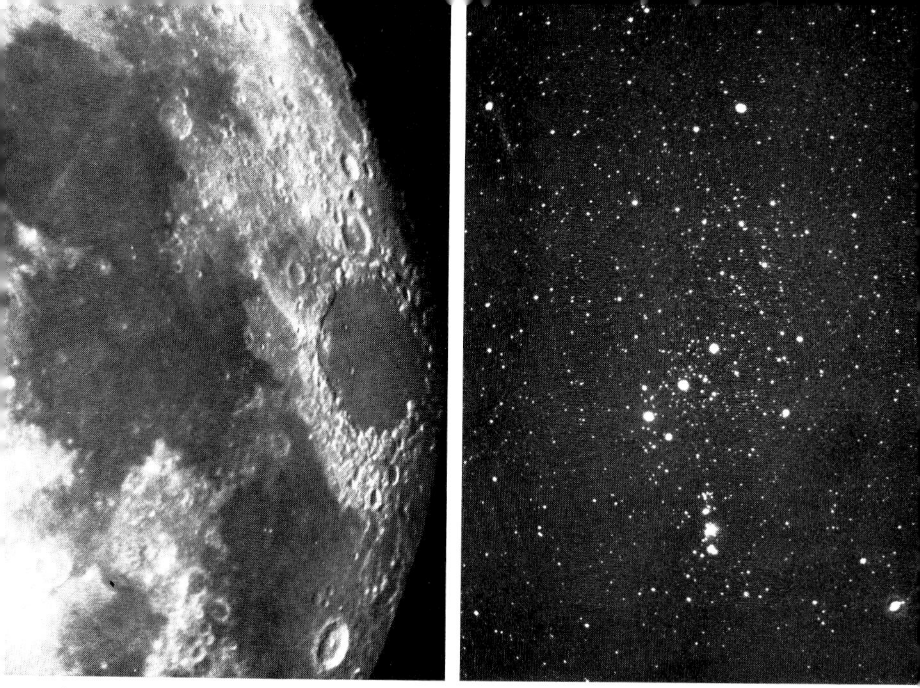

Plate 1: (above left) Part of the western side of the Moon as seen from Earth, showing the Mare Tranquilitatis (centre) where the first Apollo Moon landing took place, the Mare Serenitatis (top), Mare Foecunditatis (bottom right) with the prominent crater Langrenus, and the almost circular Mare Crisium (centre right). Maria Serenitatis and Crisium are both sites of positive gravity anomalies (mascons). The shading of the dark floors of the Maria indicate successive lava flows. *(Photograph: Richard Knox)*

Plate 2: (above right) The constellation of Orion taken with a camera tracking with the sky for 15 minutes to show stars as faint as magnitude 10. Betelgeuse (top left) is apparently less bright than Bellatrix (Gamma, top right) owing to the relative insensitivity of the photographic emulsion to the red giant's colour. Compare with the class-B8 Rigel (lower right). The Great Nebula can be clearly seen below Orion's Belt. *(Photograph: Richard Knox)*

Plate 3: (below) A close-up of the Sun's surface taken in H-alpha light to show the structure of the surface and a sunspot. *(CSIRO, Australia)*

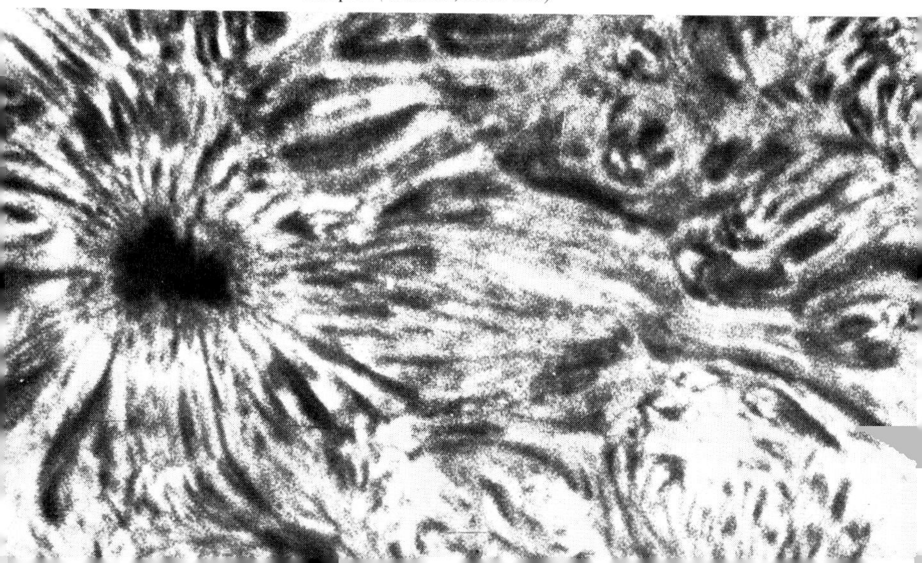

Plate 4: The surface of Mercury, photographed by Mariner 10 in March 1974. The crater near the centre with the deep internal shadow is about 12 km across. *(Photograph: NASA)*

Plate 5: The Moon, as seen by the Apollo 16 astronauts in June 1972. The Mare Crisium is near the horizon (top left). The two Mare areas below the Mare Crisium are the Mare Marginis, and Mare Smythii (near centre left), features on the horizon of the Moon's western side as seen from Earth. Other features to the right of these in the photograph are always invisible from Earth, showing how cratered features dominate this side of the Moon. *(Photograph: NASA)*

Your telescope also showed that on December 24, Europa emerged from eclipse at $2^h\ 23^m$, and by the evening of Christmas Day had swung well to the east of the planet and had begun to approach once more. Clearly, it had also begun an eclipse around noon on December 27 when invisible in the daylight.

So the *synodic period* of the satellite is half the interval December $23^d\ 22^h\ 08^m$ to $31^d\ 00^h\ 42^m$, i.e. $3^d\ 13^h\ 17^m$. (The synodic period is the apparent time the satellite takes to make a revolution of the planet as seen from Earth, and it varies slightly, so, strictly speaking, an average value of many timings should be used.)

Now from the August eclipse we can work out how many revolutions of the planet Europa has made up to December 31^d $00^h\ 42^m$. The interval is $142^d\ 3^h\ 34^m$, so Europa will have made about 40 revolutions. The expected time is $40 \times 3^d\ 13^h\ 17^m$, which comes to only $142^d\ 3^h\ 20^m$.

The apparent loss of 14 minutes is due to the fact that—as we find from our other calculations—Jupiter was at a distance of 6.14 AU from Earth in August, and 4.38 AU in December. The extra 1.764 AU in August had to be covered by the light from the satellite in a finite time, resulting in the 'error' in the actual times.

From these results, taken from actual times for the dates in question, we have found that 1.764 AU is the distance light covers in 14 minutes. This is $14 \times 60 \times 299\,792.5 =$ 251 825 280 km, so that

$$1\ \text{AU} = \frac{251\,825\,280}{1.764} = 142\,758\,095\text{km}.$$

This is within 5 per cent of the true value, 149 600 000km, the errors arising from the use of actual values rather than the average periods at the dates in question, and other small approximations.

As well as demonstrating another technique for determining distances across space, this example indicates the importance of time scales in the study of the Universe. In the case of Jupiter (on our doorstep by comparison with the stars) the light we see with our eye or telescope has taken 50 minutes to reach us when Jupiter is on the opposite side of the Sun to the Earth (when the planet would be at *conjunction*) and 33 minutes when Jupiter is closest (at *opposition*, or opposite the Sun in our sky). We can express the distance to Jupiter in terms of the time it takes light to reach us, for example 33 'light minutes' at opposition. The Sun is about 8 light minutes from Earth, and Pluto at opposition (as in 1989) is 10 light hours from Earth. But the nearest star is 4.2 light *years* away, 265 600 times as far away as the Sun.

Just an ordinary star

The Solar System, like the atom, turns out to be largely empty space, with over 99.8 per cent of the total mass of the solar system confined in one sphere 1 392 000km in diameter, the Sun. The colossal mass of the Sun, almost 2×10^{30}kg (2 followed by thirty zeros) is an essential factor in the survival of life in the Solar System, since it is this very mass which is being converted into energy.

Bearing in mind that the Sun measures about $\frac{1}{2}°$ diameter as seen from the Earth, the proportion of the energy given out by the Sun reaching the Earth is very small indeed. The total amount of energy given out by the Sun in all directions each second is 3.9×10^{20} megajoules (which is about 100 million million million kilowatt hours every second, for those who would like to imagine paying the Sun's energy bill). This was known, to within an order of magnitude or two, to astronomers of the last century who from a few quick calculations soon found that to produce that much energy by any known means, for example some chemical process such as simply burning the Sun's material, the Sun would have been exhausted long ago. Clearly, to achieve such prodigious output, an unknown method of energy production had to be involved. The final solving of the mystery yet remains to be done, although we now have a fairly good idea of the sort of thing involved.

Sir Isaac Newton did much work in the science of light and optics, as well as in mechanics, and discovered that when white light is refracted, by passing through a medium of different optical properties from those of air, it is dispersed by refraction into different colours. Actually, refraction occurs in any medium, depending upon its refractive index (an intrinsic quality) and the angle at which the light falls on the medium.

A Scotsman, Thomas Melville, observed in 1752 that if the refracted sunlight was focused from a point source, the resulting spectrum appeared irregularly bright. Then, in 1802, Dr William Wollaston found dark areas in the spectrum of the Sun, when a small part of its light was focused through a prism. This was the beginning of spectroscopy, perhaps the most important tool in the astronomer's armoury.

In 1815, a German physicist, Joseph von Fraunhofer, designed a spectroscope with which, by passing sunlight through a slit to give an easily observed and widely dispersed spectrum, he found that the Sun's spectrum was full of dark lines (the 'lines' are, in fact, the same shape as the spectroscope slit and become visible when the dark or absorption frequencies in the spectrum are not swamped by bright light, as is the case when a wider aperture is used).

Bright parts of the spectrum are due to the emission of energy by atoms of a particular element, or by groups of atoms in the molecules of chemical compounds. The bright emissions occur at particular wavelengths of light (and in the invisible regions of the spectrum). These wavelengths depend on the elements involved, and the energy states of the atoms and molecules. Similarly, gaseous molecules present between the source of radiation and the spectroscope can absorb light energy at particular wavelengths, also characteristic of the elements present in the absorbing gas. So both the bright and dark lines, called emission and absorption lines respectively, indicate the elements present in the body giving out energy, or in the medium between that body and the observer.

Spectroscopic analysis began in earnest in 1861 with a much improved spectroscope designed by Bunsen and Kirchhoff, and later, in 1882, Sir Norman Lockyer discovered a line in the solar spectrum which could not be related to any known element. This element was apparently present only on the Sun, and hence it was named helium, after the Greek *helios*, which means 'the Sun'. Helium was later found on Earth.

This was the first step towards explaining the basic process of the Sun's energy production. In the last century, Newton's gravitational laws had allowed the mass, and hence the density, of the Sun to be estimated. Despite its enormous mass, the Sun was found to be only about half as dense again as water. Clearly, as confirmed by the spectroscope, the Sun was largely composed of the lightest element of all, hydrogen.

The physics and chemistry of the nineteenth century showed that merely burning the chemical fuel of the Sun would have exhausted it within a few thousand years. If it was contracting under the tremendous force of gravity due to its enormous mass, it could shine for some 30 million years or so, but it was known that animals had been on Earth for far longer than that, so the Sun obviously had some other way of producing energy. Observation of the Sun with a telescope can be very dangerous if elementary precautions are not taken, and it is worth focusing the Sun's image on a piece of paper from a small 'magnifying glass', say a 30-mm convex lens, to see what can happen to your eye if you try to look at the Sun through any telescope, small binoculars, or other optical device.

NEVER try to observe the Sun directly with a telescope, and beware of so-called Sun filters. They have to absorb all the energy you see burning the piece of paper, and can shatter in the heat. Moreover, some filters do not remove the very harmful ultraviolet wavelengths present in the sunlight.

The Sun, however, has so much light to spare, that it is simple to project a large image into a shaded box from the eyepiece of a telescope or one ocular of a pair of binoculars.

The surface of the Sun

On most occasions, the Sun's disc will be seen to be spotted with dark spots, and closer examination will show that the entire disc is granular in appearance. These sunspots, discovered by the Chinese astronomers with the naked eye and the Arab astronomers later (also well before the telescope) are huge disturbances in the Sun's visible surface, the *photosphere*, at a lower temperature than the surrounding surface, about 3900K compared with 5700K. (The Kelvin, K, is the absolute unit of temperature, starting at –273° on the Celsius scale; thus 5700K is 5427°C.)

The lower temperature of sunspots results in a dramatic contrast in brightness so that, by the time we have reduced the brightness of the photosphere to an observable level, the sunspots appear black. But at 3900K the sunspot is still very hot indeed! Sunspots are associated with intense magnetic fields, due, it is thought, to the compression of the entire sunspot area by the denser surroundings, but this is still speculation.

It has been known since 1722, however, that sunspots have an influence on the Earth's magnetic field, and since 1851 that sunspots increase and decrease in number on the Sun's surface over a semi-regular 11-year cycle, although there is now evidence that there have been prolonged periods of inactivity during some periods of history.

The magnetic polarity of a sunspot is determined by the direction of the field emerging from the Sun, and, since sunspots appear in pairs, or more complex groupings, of opposite polarity, it is clear that the field links the spots together. Moreover, the space between spots is often low in magnetic activity, suggesting a field taking the form of an enormous arch above the Sun's surface.

Similarly, the Sun's surface can show bright patches, called *faculae*, especially near the limb (the edge of the disc) where the contrast is enhanced by the low angle at which we are observing the photosphere, and which causes 'limb darkening'. Even brighter disturbances can be seen on occasions, although really spectacular flares, such as first seen by Richard Carrington in 1859, are rare. When the Sun is observed by filtering the light to one of the wavelengths of hydrogen, many more flares can be seen. During a total eclipse of the Sun, enormous arches of plasma (gases heated until highly ionised) are sometimes seen in profile emerging from the limb of the Sun. These *prominences* are at temperatures typically of 300 000K and can reach heights above the Sun's surface approaching half a million kilometres—over a quarter of the Sun's diameter.

During solar eclipses, the Sun is seen to be surrounded by an extensive outer atmosphere, the *corona*, formed of extremely hot, tenuous gases. Only towards the beginning of the 1970s it was found that this atmosphere, at a temperature of about one million Kelvin, extends deep into the Solar System, with a 'wind' of charged particles blowing at up to 600km/s, which cause magnetic storms and aurorae when they interact with the Earth's magnetic field.

So enormous electrical and magnetic forces are at work in the Sun—but not even these can explain the output of energy.

It was not until the development of atomic physics that it became clear that the Sun's energy is derived from nuclear forces, and that in the Sun hydrogen is being transmuted by fusion into helium and a small quantity of other elements, with corresponding release of energy. So much energy is available in the nuclear processes, that the Sun can continue to convert its mass into energy, as it has already done for over four thousand million years, for another ten million million years or so, assuming it continues to produce steadily at its present rate. However, owing to changes which take place in the behaviour of stars as their nuclear furnaces begin to run down, the lifetime of the Sun is likely to be quite different from the figures just given.

What is believed to be happening in the Sun is that, deep inside where the pressure due to the Sun's enormous gravitational force is so great that temperatures reach 10 000 000K and densities are 100 times that of water, no atoms can retain their electrons, except possibly in the traces of the heavier elements which may exist. Under these tremendous temperatures and pressures, the conditions exist to fuse the bare nucleus of the hydrogen atom to another nucleus, forming the nucleus of an atom of helium, liberating energy in the process, and resulting in a loss of mass of the hydrogen nucleus of 0.7 per cent. It is this tiny loss of mass which limits the output of the Sun, since ultimately, the 'fuel' must be exhausted. Yet the Sun, losing mass at about 4 million tonnes per second, can still continue for thousands of millions of years yet.

The process by which the actual conversion of hydrogen to helium takes place is still not fully understood. Since 1939, it has been clear that two mechanisms are possible. In the first, the carbon–nitrogen cycle, protons (the positively-charged hydrogen-atom nuclei) are absorbed by carbon nuclei to form unstable nitrogen nuclei. These decay into further carbon isotopes, and each absorbs another proton to form oxygen nuclei. These decay in turn to form carbon and helium, giving a net conversion of four hydrogen nuclei into one helium nucleus.

In the second process, the proton–proton cycle, two protons collide with sufficient energy to combine to form the nucleus of deuterium, an isotope of hydrogen. A further proton joins this nucleus to form the nucleus of an isotope of helium, helium-3. These can combine in pairs to form one helium-4 nucleus, plus hydrogen. Once again, the net result is that four protons become one helium nucleus.

Since H. A. Bethe proposed these cycles in 1939, other sequences have been found which provide similar conversions of elements through the medium of other heavier elements. It had been generally accepted that the proton–proton process with the production of the isotopes beryllium-7 and boron-8 must be the process involved at the temperatures of the Sun's interior, which is too low for the carbon–nitrogen–oxygen sequence or other heavier-element sequences to take place.

Astronomy underground

The key to verifying the reactions taking place deep in the Sun, quite out of reach of almost every form of detection, was the neutrino. In the proton–proton reaction, the first stage consists of the combination of two protons to form the deuterium nucleus. At this stage, a neutrino and a positron are emitted. Now the neutrino is one of the basic sub atomic particles; the photon, which is the basic quantum of light energy forming all the light we see from the Sun and stars, is another. Whereas we need to get as high as possible in the Earth's atmosphere, or above it altogether, to catch the easily-absorbed photons, to detect the elusive neutrino we must go in the opposite direction. A neutrino telescope, far from being sited aboard a space laboratory, must be buried in a cavern a considerable distance underground!

This is because the neutrino has no charge and no mass. Its energy is due entirely to its velocity, which is that of light, and it so rarely reacts with anything that the Sun, despite its immense mass, would appear transparent to a neutrino, which can whizz through hundreds of light-years' thickness of lead as if it were not there. However, the highest-energy neutrinos react with some isotopes, very occasionally.

One such neutrino reaction is the collision with an atom of chlorine-37 (chlorine with two extra neutrons) to form argon-37 (argon with three neutrons missing). The argon-37 produced could then be detected by its radioactivity. This was the experiment set up in 1968 in the Homestake goldmine in South Dakota, one mile beneath the Earth's surface where other particles could not reach, by Raymond Davis Jr of the Brookhaven National Laboratory, USA. A 400 000-litre tank of a cleaning fluid (C_2Cl_4) was placed in the mine and monitored for argon-37. The first results were discouraging, under 3 per cent of the number of neutrinos expected being found.

The reasons for this setback could be either that the neutrinos are not produced by the round about proton–proton reaction via heavier elements, which would produce the more energetic neutrinos, or that they are produced at weaker energies, but in far greater numbers, by the basic proton–proton process. But that still does not explain the low rate of neutrino detection. Since the first experiments failed to detect enough neutrinos, the sensitivity of the experiment was stepped up.

However, the rate of neutrino capture is still only about 25 per cent of that expected.

The new Sun

In recent years, the Sun has been examined from space at wavelengths impossible to use from the Earth's surface, where these wavelengths are screened by the atmosphere. The studies at X-ray wavelengths have shown that the activity in flares is concentrated in a small core at the centre of the disturbance, forming very 'bright' (in the X-ray spectrum) spots on the Sun. Similarly, entirely unexpected 'holes' in the solar corona were discovered to exist over the poles and around adjacent regions.

Other surprises have been coming thick and fast: in 1975 it was found that the Sun pulsates, its diameter oscillating with an amplitude of some 10km every 50 minutes. Now this is less than one thousandth of one per cent, so it needed extremely accurate detection. Two groups of astronomers also claim to have found longer-period oscillations, about 2 hours 40 minutes, although this finding has been challenged. This is similar in principle to the short-period small-amplitude variations of brightness found in many stars. It may also be related to the much more obvious variations in brightness of some of the well-known variable stars, which we will examine later.

The Sun is a slow nuclear reactor (it is *not* like the thermonuclear bomb, as so often claimed, any more than is a nuclear power station; and neither the Sun nor the power station can explode, which requires the very rapid conversion of one element to another). That these solar reactions are slow is fortunate for us, as it gives the Sun another 5 000 000 000 years or so in which to exhaust its hydrogen supply. Then some dramatic changes will take place, as we will see later. But in the meantime the Sun is literally pulsing with energy. The pressure of this energy from inside the Sun is balanced by the enormous inward pressure due to the Sun's gravity, so perhaps these short-period pulsations will provide a clue to the nature of the reactions deep within the Sun.

The debris called 'the planets'

With 99.8 per cent of all the matter in the Solar System concentrated in the Sun, the 0.000 6 per cent that is used up in making the four inner planets, Mercury, Venus, Earth and Mars, and their satellites, seems hardly worth more than a paragraph. The entire mass of the planets, asteroids, comets, dust and any remaining matter amounts to 0.135 per cent of the Sun's mass, and 70 per cent of the total planetary mass is Jupiter's—which perhaps deserves two paragraphs.

In fact, of course, it is important that the planets should be only very small. If a planetary body is large enough, its own force of gravity would cause it to collapse. This process would continue until nuclear reactions triggered by the enormous pressures and temperatures created by the collapse produced enough outward pressure to balance the gravitational forces. The planet would then be a star in its own right, and would form a binary system with its primary star. Many examples of binary systems exist in various stages of development, suggesting to us that this is what in fact happens to many stars and their attendant bodies. It is one of the reasons we can be confident that other planetary systems exist.

But in our case, had Jupiter been large enough to become a star, it would have so disturbed the orbits of the other planets that in all probability they would have long ago ceased to exist, or would have been too far away from the Sun and its companion to receive the energy essential for life.

It would also be most uncomfortable for life as we know it on a planet with a gravitational force at the surface much different from our own. If it were much smaller, such as Mercury (4880km diameter, 0.38 Earth-diameters) or the Moon (3476km diameter, 0.27 Earth-diameters) or even Mars (6787km diameter, 0.53 Earth-diameters) the planet's atmosphere would have become thin or disappeared altogether, so that at best only very rudimentary life forms could survive. There is virtually no chance of finding life on Mercury, which, like the Moon, lost its atmosphere into space early in its history; and the *Viking* landers on Mars have shown that there is little chance of even simple forms of life being found there. But, bearing in mind that microscopic *streptococcus* bacteria originating on Earth survived over two years on the *Surveyor 3* spacecraft on the Moon, before being returned to the Earth in the craft's television camera by the *Apollo 12* crew, there is still a possibility that a tiny foothold of life may yet be found somewhere in the Solar System.

If the surface gravitation were much greater than Earth's, the atmosphere would contain many lighter elements and their compounds which had never been able to escape the gravitational attraction of the planet as happened on the 'Earth-type' planets at an early stage of their history. The only significant example in the Solar System is Jupiter, with a surface gravity over 2½ times that of Earth (although the 'surface' of Jupiter is believed to be liquid hydrogen). Moreover, the high surface gravity would bind the atmosphere tightly to the planet, resulting in enormous atmospheric pressures and the exclusion of heat, light and gases such as oxygen from the surface. If we loosen our imagination a bit, of course, we can hope (as some have suggested) for living organisms floating high in the thick atmosphere of a planet such as Jupiter, but this is pure speculation.

As discussed later on, there is probably an enormous number of life-bearing worlds, but the galaxy that we inhabit is so vast that even if life occurs on most stars of the same type as our Sun, *and* has taken a similar timespan to evolve since the Universe began (the Sun is 'only' some 4600 million years old, compared with about 19 000 ($\pm$ 5000) million years—the current estimate of the age of the Universe) it is still unlikely that intelligent life at a state of evolution comparable with that on Earth will occur within tens of light years of Earth.

The main features of the planets of the Solar System are summarised in the scale diagram on page 30. Apart from its position, each planet has unique features, and there are also many features in common. There is much still to be discovered. The *Mariner 9* mission to Mars increased our knowledge of the planet 10 000-fold in terms of photographic resolution, even including the three previous Mars *Mariner* missions, according to an estimate by Carl Sagan, professor of astronomy and space sciences at Cornell University. Since *Mariner 9* there have been the two *Viking* landers on the planet. Each new discovery made by spacecraft in the Solar System has shown how little was known previously—but equally, many new problems have been found.

Mercury, the elusive planet

A good example of how 'facts' have recently been upset by new techniques is the simple case of Mercury's period of axial rotation. This was quite categorically given as 88 days in respectable works of reference as recently as 14 years ago. Mercury is an elusive planet, and difficult to observe from Earth-based telescopes, as it is never more than $27\frac{3}{4}°$ from the Sun (called its *elongation*), about the same angle as a 300-mm rule held at arm's length makes to the eye. This means that most observations must be made while the Sun is above the horizon, since otherwise the planet is low in the sky and the elusive details are lost owing to the effect of the increased thickness of the Earth's atmosphere when observing any body thus situated.

Earth-based observations of surface markings on Mercury gave the half-expected result that the planet rotated once on its axis as it made one revolution about the Sun in its orbit, thus keeping one hemisphere turned always towards the Sun. This 'captured' rotation of small bodies close to large primary bodies is believed to be common. It occurs in the case of the Moon, for example, and is due to tidal forces set up by the primary body on small unevenly-distributed masses in the smaller body, causing its axial rotation to slow down until synchronised with its period of revolution about the primary.

In the case of Mercury, it was discovered in 1962 by radio astronomers from the University of Michigan that thermal emissions from the supposedly 'dark' side of Mercury were too high for a part of the Solar System for ever hidden from the Sun; and then, in 1965, radar pulses reflected off features on the limb of the planet from the giant Arecibo radio telescope revealed that the axial rotation period was about 59 days. The significance of this was recognised by the Italian physicist Giuseppe Colombo, who suggested that the figure would be exactly two-thirds of the *sidereal* period (the time for one revolution about the Sun) of 87.969 days, giving 58.65 days for the axial rotation period. So Mercury was indeed 'captured' by the Sun, in that it rotates exactly three times while orbiting the Sun twice, suggesting that the tidal action of the Sun on Mercury has trapped it at a particular resonant 'harmonic' of its frequency of rotation.

In March 1974, *Mariner 10* flew by Mercury, went on to fly past Venus, and returned to Mercury in September. It returned again in March 1975, and its orbit continues to take it near to the planet once every 176 days. These encounters have enabled the rotation period of Mercury to be ascertained precisely: it is exactly as predicted by Colombo.

Many astronomers (and science-fiction writers) predicted that Mercury would be like the Moon, and they were not disappointed. The extreme temperatures, topping 400°C in the day and dropping to about –170°C at night (when the heat is quickly radiated from the surface unshrouded by any atmosphere) are similar in their contrast to those on the Moon, where daytime and night-time extremes are 130°C and –190°C. The maximum temperatures on Mercury are, of course, much higher because of the proximity of the Sun.

The cratered surface is also very similar to the Moon, including the distinctive white 'splashes' of the ray craters (see photographs on pages 33 and 34), but there are important differences. The large, flat, dark plains, or 'seas', of the Moon are not seen on Mercury, and Mercury has craters which are less densely packed, with relatively extensive barren areas in the 'highlands' (to use the lunar terminology).

Perhaps the most surprising difference, however, was the detection of a relatively strong magnetic field, about one-tenth as strong as the Earth's and aligned along its axis of spin, as is the Earth's. Neither the Moon, nor Mars, nor Venus has an appreciable magnetic field, and its presence in such a small body as Mercury is not easily explained in the same way as the Earth's magnetic field, which is usually said to be generated by currents in the Earth's iron core as the planet spins rapidly on its axis. The *Apollo* landings on the Moon have shown, as expected, that the interior of the Moon is almost dead, but clearly in the past vast volcanic disturbances (probably triggered by the impact of a large meteorite early in the Moon's history) caused great upwellings of lava across the 'seas'. But Mercury's surface shows little sign of volcanic activity, so the interior of Mercury is probably at least as dead as the Moon's. Mars and the Moon have similar low densities, suggesting iron cores much smaller than the Earth's and Mercury's. Mars also shows clear evidence of primaeval volcanic activity. As we shall see later, there is also evidence of continuing, though rare, volcanic activity on the Moon manifesting itself on the surface as well as deep down, but the Moon has virtually no magnetic field now, if ever it did. Making sense of these various data is difficult, and perhaps the magnetic field of Mercury has quite a different explanation.

Mercury was observed by the most ancient astronomers, although it was given two names by the Egyptian and by the early Greek astronomers, depending upon whether the planet was seen in the morning or the evening. When Mercury is to the east of the Sun, it rises and sets after the Sun and is termed the evening star. When the planet is to the west of the Sun, the reverse applies and it is called the morning star. The evening star was called Horus by the Egyptians and Hermes by the Greeks, whereas the morning star was respectively called Set and Apollo.

The difficulty of finding the elusive planet, and the rapid change of its position in the sky relative to the Sun (due to its short sidereal period) over the few days in which it is easiest to spot, gave rise, no doubt, to its being associated with the messenger of the gods. Some say that Copernicus never saw Mercury, but this is probably not the case, even though he was much more concerned with the theory than with practical observation.

The best time to find the planet as a morning star is close to western elongations in September (for observation from the northern hemisphere of the Earth) and as an evening star about March. For southern-sky observers, the reverse applies. The reason for this is that, at the times of day and year mentioned, the path of the Sun across the celestial sphere, called the *ecliptic*, which the planets also follow closely, is most steeply inclined to the horizon at sunrise or sunset. This means that Mercury is as high as possible above the horizon after sunset or before sunrise. It helps to use binoculars, so long as the Sun has set or has not yet risen, and if one of the brighter planets (such as Venus) is also in the sky, it is easy to scan along the line between the estimated position of the Sun (below the horizon) and the brighter planet.

Venus: the hostile world

A celestial Garden of Eden, a beautiful water-covered planet with floating tropical islands, was the idyllic picture that C. S. Lewis painted of the second planet from the Sun, Venus, in his book *Perelandra* (or *Voyage to Venus*). In general, this is how many science-fiction writers saw the 'Bringer of Peace', the planet that the ancient astronomers had named after the goddess of love.

The ancient astronomers aptly named Jupiter after the king of the gods, but it could hardly have been more inappropriate when they named Venus. The planet Venus is Earth's nearest 'independent' neighbour, and was, until quite recently, favoured as one of the best hopes for life in the Solar System apart from the Earth. But when the Russians made the remarkable Venus landings with the *Venera* space probes, and when surveys from space were made by the *Mariner 10* spacecraft (the same one that made the spectacular Mercury encounters), it became clear that Venus was closer to a hell than a paradise.

The surface of the planet is a rocky desert, like most of the inner planets, but in Venus' case it is composed of rocks eroded by surface temperatures of 750K (about 480°C), high enough to expel vast quantities of carbon dioxide from the surface minerals and so to produce an atmosphere so heavy with this gas that the surface pressure is 90 times that at the Earth's surface.

At high atmospheric altitudes, the carbon dioxide is bombarded with intense ultraviolet radiation from the Sun, which in Venus' sky appears 40 per cent larger than on Earth. This intense radiation produces carbon monoxide and ozone which react with the hydrogen ions of the solar wind to form water, and with other compounds also expelled from the surface rocks to form complex and nasty substances such as sulphuric acid with traces of hydrochloric and hydroflouric acids.

These compounds react with particles and chemical compounds in the atmosphere in complex cycles which result in dense unbroken clouds over the whole planet. It is these clouds which have hidden Venus' secrets for so long from Earth-based observers, have given the planet its brilliant yellow–white featureless appearance in a telescope, and have made it the brightest object in the sky after the Sun and the Moon. It was the formation of clouds early in Venus' history that brought about the amazing atmosphere of the planet.

The proximity of the Sun, it is believed, converted the atmosphere from something similar to the Earth's at the same stage of evolution to its present hostile form. The water in the planet's atmosphere and on the surface was quickly vaporised in the strong sunlight, causing clouds of water vapour to begin shrouding the planet. Clouds act as a heat insulator, as is commonly demonstrated on Earth when the clouds clear suddenly at evening and at once the temperature drops. This is because the clouds largely reflect back the heat radiated by the surface rocks, which are efficient thermal-storage materials. (Some heat is nonetheless lost by radiation and convection through the clouds.) The principle is the same as in a greenhouse, and

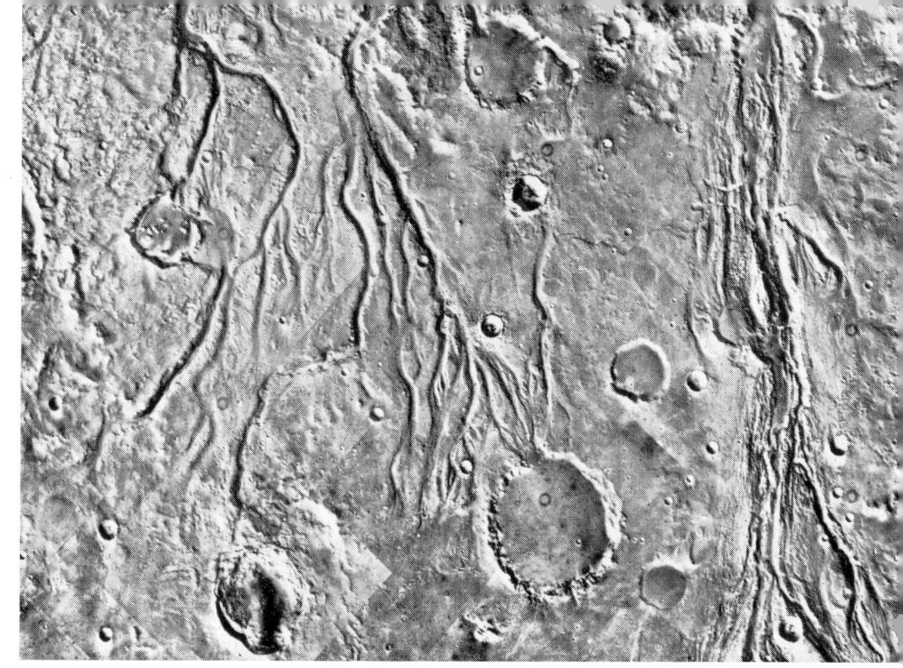

Plate 6: The area to the east of the Lunae Planum plains of Mars, photographed from the orbiter part of Viking. The surface of the planet slopes towards the top of the area shown. Flooding from the Lunae Planum could have caused the channels that in some places cut through older creaters. *(Photograph: NASA)*

Plate 7: A mosaic of photographs taken from the Viking orbiter at 480 km from Phobos, Mars' inner satellite. Phobos is about 21 km across and 19 km top to bottom as shown here. Note the strikingly linear crater chains and ridges. The large crater seen obliquely near the centre bottom is named Hall after Asaph Hall, discoverer of the Martian satellites. This crater is 5 km across and contains the satellite's south pole. *(Photograph: NASA)*

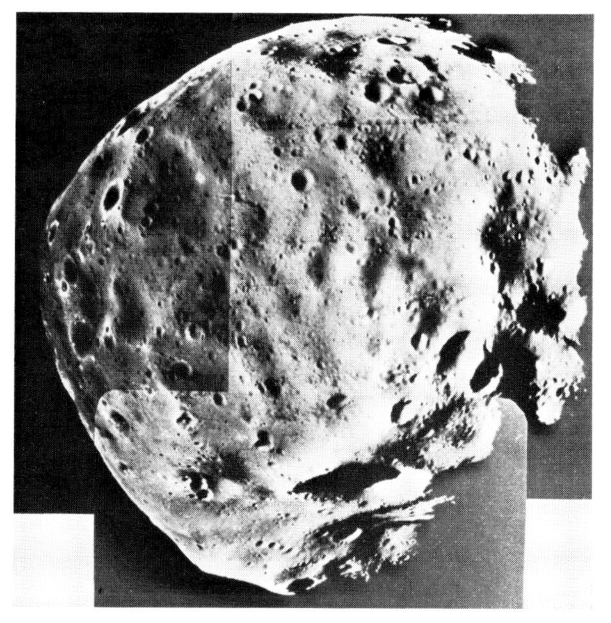

Plate 8: Jupiter's Great Red Spot taken in blue light from the Pioneer 11 spacecraft from 545000 km. The feature is large enough to contain three Earths! Surface details (if Jupiter's thick atmosphere can be called its surface) show the white oval beneath the Great Red Spot (G.R.S.), one of three usually 120° apart surrounding the planet at this latitude. The 'eye' in the centre of this oval suggests rotation, like a cyclone. The streams of cloud north and south of the G.R.S. flow in opposite directions. *(Photograph: NASA)*

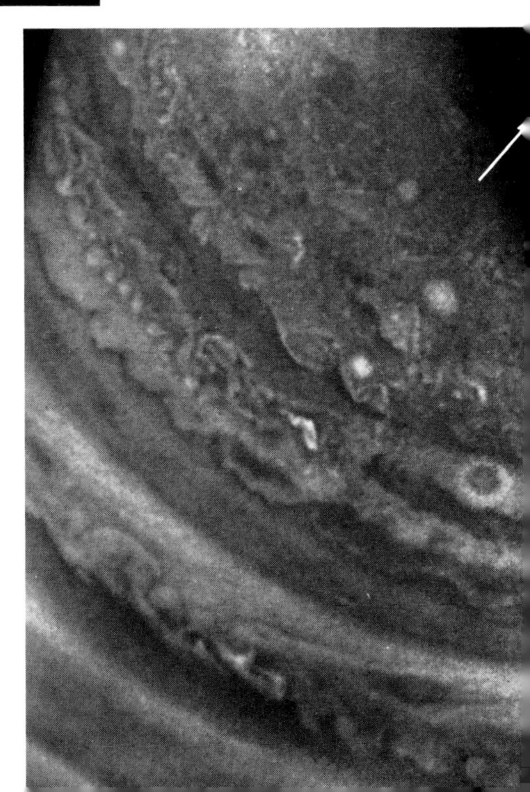

Plate 9: Part of Jupiter's north temperate zone seen from 600000 km in blue light from Pioneer 11. The area near the north pole appears to contain disorganised hurricane-like storms. The effects of the counter-flowing jet streams between the different belts is to produce sharply-defined spiral features and scallops, causing wind speeds at the shear points between them of up to 500 km/h. *(Photograph: NASA)*

'the greenhouse effect' is the term commonly used to describe this process, thought to have formed and maintained Venus' atmosphere.

As the water-vapour clouds caused the surface temperature to rise, eventually the boiling point of water was reached causing a massive increase in the thickness of the cloud cover, and a corresponding further rise in temperature. This continued until carbon dioxide began to be liberated from the carbonates in the surface rocks, and the present atmospheric chemistry began.

However, there are plenty of puzzles left: if the Earth and Venus had common origins, then Venus might be expected to have a plentiful supply of water, like the Earth. But although the greenhouse effect could occur on the Earth if a dramatic increase in the surface temperature were to take place, the buildup of cloud from the much larger quantity of water on the Earth could not result in a completely dry atmosphere such as Venus now has. This suggests that Venus started life with only very little water.

The slow rotator

Special techniques have recently been developed to obtain high resolution radar maps of the planet's surface, which is impressive bearing in mind that the disc is only about one minute of arc (1 arc min = $\frac{1}{60}$ of a degree) in angular size when at its closest to Earth.

Until these experiments had been carried out, one of the simplest data about the planet, its period of axial rotation, had been a mystery. The clouds prevented any observation of surface markings, but measurements in 1956 by J. D. Kraus of the radio emissions from the planet showed a periodic variation of about 22 hours 17 minutes. Since both the Earth and Mars have rotational periods of about 24 hours, this seemed a likely value, and was widely quoted. However, by the mid 1960s it became clear that many astronomers were finding their results incompatible with a rotation period of about one Earth-day.

One surprising result of a radio observation by Bernard Guinot was that Venus might even rotate towards the west, the opposite direction to that of all the other planets (except Uranus, which will be mentioned later). This retrograde axial rotation was later confirmed with the radar measurements, and was found to be once in only 243 days.

The radar observations also indicated the presence of large circular formations on Venus like the large craters found on all the inner planets, but, like the Earth, the formations are less well marked than on the other planets because of erosion by the planet's atmosphere.

Venus has virtually no magnetic field. But this is consistent with present theories of how planetary magnetic fields are caused by the currents set up in the liquid core due to the rapid rotation of the planet, since Venus rotates so slowly. But if it had had a magnetic shield to protect it from the solar wind (the ionised hydrogen pouring from the Sun) the effects on the atmosphere would have been quite different; who knows what this might have meant in terms of the evolution of the planet and the possibility of life on its surface?

Venus, more than any other planet, has sprung the biggest surprises and posed the most awkward questions of planetary evolution in recent years. Its terrible surface conditions make the very survival of a space probe extremely difficult to achieve. There are, no doubt, more shocks in store.

Venus is the easiest of any planets to identify since it is by far the brightest. However, it is not always conveniently placed. It has a maximum elongation of 47° from the Sun, and when this occurs at the favourable inclinations of the ecliptic as described for Mercury, the planet is so bright in the evening or morning sky (nearly 12 times brighter than the brightest star, Sirius) that it is seen long before sunset by many who know little of astronomy, who then may be convinced they have seen a flying saucer!

In fact if you look carefully, with something to guide your eyes (and as usual take steps to avoid looking at the Sun) you can find Venus in broad daylight so long as it is at a modest angular distance from the Sun. A simple way of observing Venus when the planet is to the east of the Sun (an evening star) is to set some low-power binoculars on a tripod to point at the Sun NOT BY LOOKING THROUGH THEM, but by casting the Sun's image on a screen (or just your hand) behind one eyepiece. (If you leave your hand there for a while, you will find out why you must not put your eye to the eyepiece.) You must find Venus' elongation from an almanac, or by calculation, and then allow four minutes of time for each degree. The resultant interval will allow the Earth to turn until Venus is in the field of view. By this time, the Sun is well out of the way, and you can look through the binoculars. You may have to move them slightly, as Venus is not exactly on the ecliptic always, but it will not be far out.

Venus is seen as a gibbous or crescent 'phase', since it is always at an acute angle to the Sun as seen from Earth. This also applies to Mercury. All the other bodies of the Solar System can reach 'opposition', when they are opposite the Sun in the sky, and due south at midnight (from the northern hemisphere). The crescent of Venus is easily seen in binoculars, and some claim to be able to detect the crescent with the naked eye. Since the human eye has a limit of resolution for the separation of two contrasting points of the order of 0.4 minutes of arc, and since observing the *shape* of a crescent of about that size would be much more difficult, such visual faculties must be exceptional.

The planet Earth

Living on its surface, we easily forget that the Earth is a planet, just as we tend to forget that the nearest star is the Sun. As we saw earlier, it is easy to prove that the world is round and not flat as most early civilisations believed. The view of Earth from space has added a new perspective to our appreciation of its beauty, and even more important, its smallness.

Planet Earth shares many characteristics with its close neighbours Mercury, Venus, Moon and Mars: it is of comparable size, density and composition, and it is reasonable therefore to conclude that it has been formed by similar processes, and in the same regions of space as the other planets of the inner Solar System. Since that formation, 4600 million years ago, the history of these five bodies has been increasingly different, giving rise to the dramatic differences in their present states.

All these inner planets were heavily scarred by bombardment early in their history when the Solar System was still full of the debris of its formation. Those scars clearly remain on the Moon, Mercury and Mars, and there is evidence that large craters also exist on Venus, although its atmospheric conditions have eroded many of the easily visible signs as on Earth. One of the most important contributions to the change in the Earth's appearance was the formation of the oceans which now cover some 80 per cent of its surface.

The formation of the Earth, it is now thought, began simply by the gravitational accumulation of debris from the material surrounding the young Sun until, under the pressure of this accumulating mass, the incident heat from the Sun and the heat generated by radioactive decay, the temperature of the inner material rose until it became molten. At this point, the denser materials sank into the centre of the Earth to form the molten core, chiefly of iron; and the lighter materials formed into a crust of slag. Fortunately, some of the heavier elements were able to reach the surface in compounds less dense than the core. Nonetheless, this is why materials such as gold, lead, uranium and other heavy materials are comparatively rare at the Earth's surface, while silicon (rocks and sand) abounds.

The age of rocks can be deduced from the traces of various radioactive elements remaining in them, since these decay into other elements over very different periods of time. The oldest rocks so far found on Earth are 3700 million years old, about 1000 million years younger than the Solar System is thought to be.

As the slag of crustal materials cooled, buckling and splitting all the while to form the primaeval continents, the mountains,

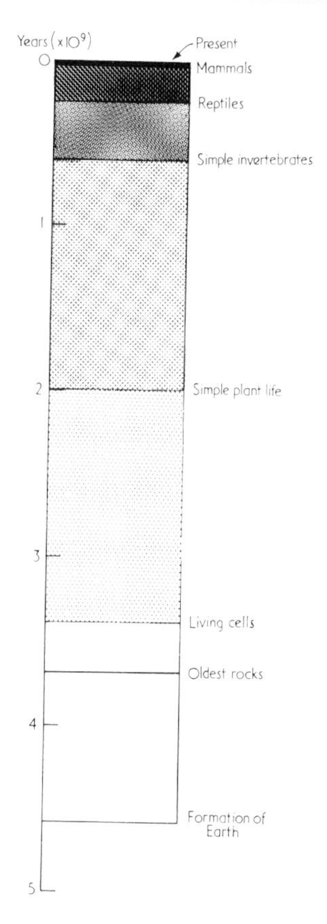

11 The approximate history of the Earth. The shaded area represents the time life has existed on Earth, the darker the tint, the more complex the lifeforms. The black line on top represents the time mammals have existed.

and the ocean beds, gases were driven out of the interior to form an atmosphere heavy in carbon dioxide and water, together with gases such as ammonia, methane and sulphur compounds. The conditions were just right, however, to prevent the runaway temperature effects which probably happened on Venus, so that the water vapour could condense freely, dissolving the heavy gases and over the centuries bringing them to the surface in torrents of rain. Here they formed other chemical compounds and were slowly purged from the atmosphere.

At the same time, the dense primordial atmosphere was being churned in the most violent storms, due to convection brought about by the heat of the Sun and the surface of the planet itself. This, and the torrential rain, would have led to electric storms of a violence scarcely conceivable now. But it now appears from laboratory experiments with electrical discharges in similar atmospheres that the first primitive organic compounds were formed in this phase of the Earth's development, and that these primaeval storms were probably the beginnings of life on Earth. The earliest-known living organisms appeared about 3500 million years ago, not much younger than the oldest-known terrestrial rocks, but it was not until 3000 million years later that advanced animal life emerged on to the land.

If we call the lifetime of the Earth to date one day of 24 hours, the oldest rocks were formed at 4.40 a.m., and life emerged at about 6.15 in the morning. But it was 10.40 in the evening before advanced life emerged on to the land, and some form of man-like creature walked in at nine minutes to midnight, followed by *Homo sapiens* at four seconds to midnight. The great civilisations began, on this scale, less than a tenth of a second before the present!

The continents themselves are only thin layers on the surface of the planet, and even now are in motion relative to each other. The crust and upper mantle of the Earth, termed the *lithosphere*, is sliding on the plastic layers of the mantle beneath, the edges of parts of the lithosphere sliding under other edges or emerging from the mantle. Thus the lithosphere is divided into *plates*, along the edges of which occur the main violent geological disturbances on Earth such as earthquakes and volcanos. The continents are slowly drifting relative to one another as the plates move beneath them, and geological and biological evidence shows that the continents were once joined in very different configurations. The easiest example can be seen on any map of the world by looking at the continents of Africa and South America which fit almost like jigsaw-puzzle pieces.

Similar processes may well be at work on Mars and Venus. As we shall see, there are major chains of giant volcanos and massive canyons on Mars, suggesting that a similar process could have taken place there, and perhaps still does.

The new Moon

When the *Apollo 8* crew, Borman, Lovell and Anders, initiated the departure from their orbit around the Earth to travel around the Moon during the Christmas of 1968 this was in many ways a much more significant 'small step' than Neil Armstrong's 'giant leap' on to the Moon's surface. This was the first occasion on which men had severed their connection with the Earth. The trajectory of their spacecraft was such that they could not return until they had made at least one orbit of the Moon. They had taken an irrevocable step into the unknown.

But even the trip to the Moon did not allow the astronauts to escape the influence of the Earth, since the Moon itself is bound to the Earth and shares a common orbit round the Sun. The effect of the gravitational attraction of the Earth and Moon on each other is to cause them to wobble about a common centre of gravity in their yearly motion around the Sun. If the Earth were invisible to a Martian astronomer, for some reason (if it were a black hole?), and the Moon remained visible, the Martian would observe that the Moon was in orbit about the Sun, but had a small periodic disturbance in this orbit with a magnitude of 0.25 per cent of its distance from the Sun and an interval of just under 27 Martian days (he would not, of course, be aware of Earth days); and, if the Moon's orbit were plotted to scale, he would see that at all times it is curving towards the Sun.

But the real significance of Neil Armstrong's 'giant leap for Mankind' was in the knowledge that he and his companions Aldrin and Collins brought back with them to Earth from the

Sea of Tranquility, locked up in a few kilograms of rocks.

Until that moment, Earth-bound scientists had had no way of examining samples of material from outer space, such as those brought to Earth in meteorites, with the knowledge of where they had been formed. With samples of Moon rock, it was possible for the first time to test certain aspects of the theories of the formation of the Solar System. Once again, the results were very surprising.

The oldest rocks found on the Earth are about 3700 million years old, but the oldest samples brought back from the Moon were some 4500 million years old, almost as old as the Solar System itself is reckoned to be. All the Earth-type planets (Mercury, Venus, Earth, Moon, Mars and possibly the satellites of the outer planets) probably began to form by a gradual accretion of mineral particles under their mutual gravitational attraction. At the same time, gases remaining in the surrounding space would have begun to condense on the newly-forming solid bodies, so that their final form would contain a mixture of the least-volatile materials (which would have been the first particles to become solid as the gases of the Solar System began to condense) and of the more-volatile substances (which would have condensed much later, as temperatures began to drop on the surfaces of the solid bodies).

Three theories of how the Earth acquired such a satellite as the Moon were widely held before the manned Moon landings, although two of these were already presenting obvious difficulties.

One theory held that the Moon somehow broke away from the Earth. If this happened early enough in the formation of the Earth, the differences in their composition could have arisen afterwards, but it is most unlikely that the Moon 'broke off' the Earth leaving the hole which became the Pacific Ocean, as was held by many people in the past. But if the Moon had separated from the proto-Earth, it would have needed a very high rate of spin to cause such an event, and the angular momentum of the present Earth–Moon system is much too small to have allowed that. Since, according to Newton's laws, the velocity of a body cannot change unless a force is applied to it (and velocity is not merely speed, it is a vector quantity, which means it has direction as well as magnitude), and since momentum is the product of mass and velocity, any such reduction in the angular momentum of the Earth–Moon system as is required by this theory implies a considerable slowing down due—presumably—to some unimaginable force, since the system is most unlikely to have acquired an enormous extra mass since the formation of the Solar System.

Another popular theory was that the Moon was formed elsewhere in the Solar System, and was later captured gravitationally during close encounters with the Earth. This is certainly a reasonable assumption in the case of many of the satellites of the outer planets, many of which rotate about their primary planets the 'wrong way' (that is, from east to west) or in orbits which are unusual in inclination to the plane of the Solar System or in eccentricity. The small planet Pluto is so abnormal in these respects that it is commonly supposed that it is an escaped satellite of Neptune. But although the Moon's orbit is elliptical, like all orbits, it is not particularly eccentric, nor is its 5° inclination to the ecliptic particularly unusual. In addition, very peculiar circumstances would be required to account satisfactorily for such a capture of a body as large as the Moon by one as small as Earth.

So the most likely explanation was that the Moon was formed out of the same basic material as the Earth in the same general region of space, and was therefore formed as a satellite.

But the Earth and Moon, it was found, were composed of significantly different proportions of high- and low-condensation-temperature minerals, the Moon being lacking in the more volatile (low-condensation-temperature) minerals.

The higher proportions of high-condensation-temperature materials in the samples of Moon rock caused quite a surprise, since if the Moon was of the same ancestry as Earth it would be expected to have the same general composition. A combination of particle accretion and simultaneous condensation of materials from the gaseous state would take place at different rates according to the respective rates of change of temperature of the two bodies, and this is probably the explanation of why the Moon has the higher proportion of high-temperature rocks and of the oldest ones. The Moon, it appears, became a solid body which we would probably have recognised easily, had we been able to go back in time to an era when the Earth was still a mass of liquid rock. Most of the incredible scarring of the Moon's surface would have taken place when there was plenty of debris still left in space while the planets were being formed, although the great lava plains of the Moon, which are fancifully (but most misleadingly) named 'seas', indicate that at that time the interior of the Moon was still very active.

The Moon keeps one hemisphere turned towards the Earth as it rotates about it, because its period of axial rotation is the same as its mean sidereal period. This is due to gravitational interaction between the two bodies causing tidal forces which over the millennia have brought the Earth–Moon system into equilibrium. The far side of the Moon was virtually unknown before space probes had orbited the Moon. All that had been seen previously from Earth was a little of the far side, exposed to Earth-based observers because the Moon's sidereal period varies slightly owing to the eccentricity of the Moon's orbit and other effects. It was therefore a complete surprise when the Russian *Lunik 3* space probe revealed that the far side had virtually no *maria* ('seas'), certainly none resembling the great maria which form the familiar 'Man in the Moon'.

Gravity anomalies

When the lunar orbiters began sending their incredibly detailed pictures of the Moon's surface from different altitudes and angles, the results were breathtaking. However, a less spectacular but much stranger discovery was that space probes in close orbits round the Moon were being slightly disturbed by variations in the gravitational attraction of the Moon. These 'gravitational anomalies' were both positive and negative: in some cases the attraction was greater than average, and sometimes less. The positive anomalies, or concentrations of mass, (*mascons* for short) occurred on both the Earth-facing and the far side of the Moon, but five of the most sharply defined mascons were found to be situated in or near the huge circular lava plains, the *Maria Nectaris, Crisium, Humorum, Serenitatis,* and, largest of all, *Imbrium.*

Why should these huge dark plains occur all on one side of the Moon, and what is the connection with the mascons? Although the largest mascons on the Earth-facing side are beneath the low-lying plains, the largest positive anomaly is on the far side, almost beneath the huge walled crater Hertzsprung.

It has been calculated by A. J. Ferrari of the California Institute of Technology that the average gravity anomalies on the near and far sides are unequal, and the average gravitational attraction is greater on the far side of the Moon, despite the greater number of mascons on the near side. He suggests that the average height of features above the mean level on the Moon is greater on the far side, so that it appears as if the Moon's crust is thicker there. This would explain why the dark basaltic material of the low-lying plains occurs on one side of the Moon rather than the other, since the lava would more readily fill the low-lying plains. The main negative anomalies may be craters unfilled by the lava flows.

Are the mascons caused only by the concentrations of lava, or are they the remains of dense planetoids buried just below the Moon's surface where they struck so long ago? The Moon's thick mantle may easily have supported a dense body below the surface which, on the Earth would have sunk through the mantle into the core of the planet. Such bodies must have bombarded all the planets at that time in the evolution of the Solar System, but the subsequent history of the craters or other scars of impact were very different on the various inner planets.

The dark *maria* of the Moon show that lava flows occurred many times in the same basins. These eruptions continued, it is believed, for thousands of millions of years after the bom-

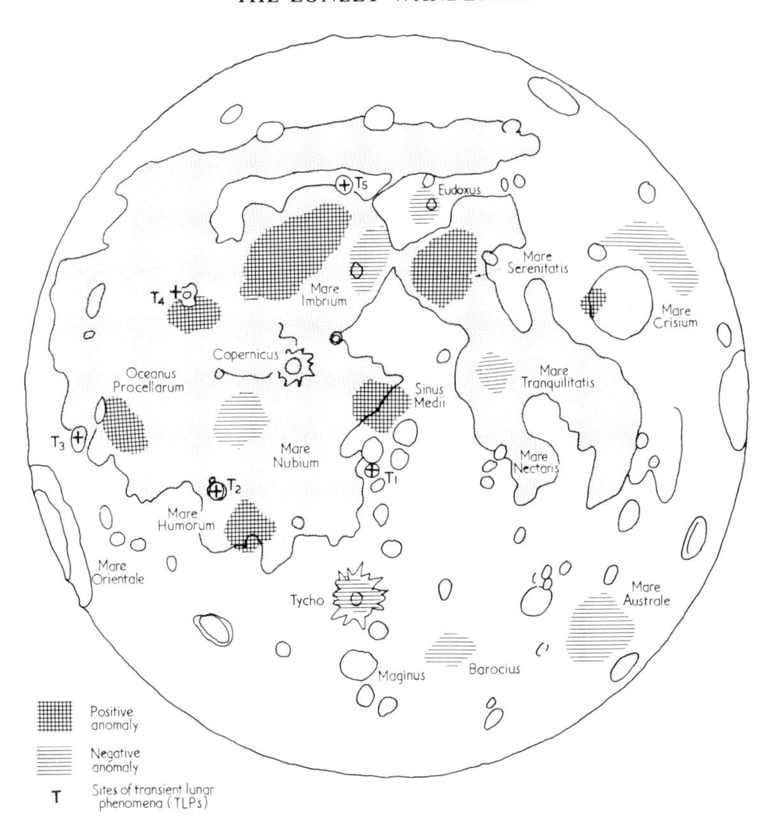

12 and 13 Sketches of the near and far sides of the Moon, showing the centres of the principal gravity anomalies (positive and negative) and the sites of the most frequent transient lunar phenomena.

bardment of the 'planetesimals' had ceased, when the gravitational forces of the planets and the Sun had almost cleaned up the remaining debris left from the early ages of the Solar System. Successive lava flows across the plains are clearly seen in a small telescope, the darker material being the most recent. The amount of cratering in the seas is clearly much less than over the surface in general, and is least of all in the darkest, most recently formed *maria*.

The data on the Moon rocks, and what they have taught us,

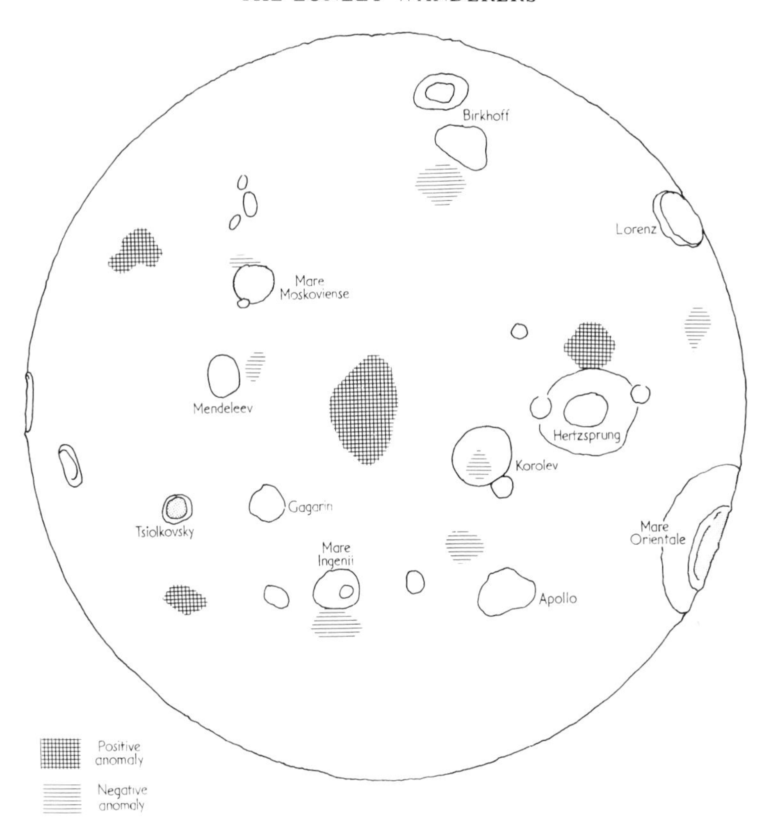

would fill several books. No terrestrial rocks have been so extensively examined! But in general they have shown that the Moon cooled rapidly to become a dead world. The seismic activity remaining on the Moon, studied for many years since the departure of the Apollo astronauts by the instruments they left behind, was found to be very small. (These instruments were switched off in the autumn of 1977.) But owing to the rigid nature of the Moon, any disturbance—such as a small moonquake (there *are* only small ones), or the impact of a body such as the lunar excursion modules, which were deliberately crashed on to the surface after the astronauts' departure—causes the Moon to vibrate for hours afterwards, rather like a bell ringing. The Moon has a rigid lithosphere some

1000km deep which now prevents any movement at the surface that could be seen from Earth. All that is thought to be left of the hot part of the Moon is a small semi-molten core 1500km in diameter.

It is possible, however, that some gases may still escape from the deep interior to the outer surface from where, like any atmosphere the Moon once had, they soon escape into space. The Earth-based observers of the Moon have on comparatively rare occasions seen hazy patches appear on the Moon, sometimes of a reddish hue suggesting dust being blown out of the crack from which the gases are emerging. Detection of these 'transient lunar phenomena' (t.l.p.) is an important job for amateur observers, since only by watching the Moon continuously with many eyes and cameras is there any chance of spotting a t.l.p. The leading amateur astronomical associations in various parts of the world co-ordinate this work.

The Martians are coming!

'Nor was it generally understood that since Mars is older than our earth, with scarcely a quarter of the superficial area, and remoter from the sun, it necessarily follows that it is not only more distant from life's beginning but nearer its end.'

These words of H. G. Wells appeared in *Pearson's Magazine* of April 1897 in his introduction to *The War of the Worlds*. Only twenty years previously the Italian astronomer Schiaparelli had described in detail how the surface of Mars was marked with extensive straight lines which he thought were channels (*canali*). According to Percival Lowell, the American astronomer who founded the Flagstaff observatory, these must be the planet-wide irrigation canals of a race of Martians fighting the death of their planet by bringing water from the melting polar caps. The planet Mars had become well established as the best bet for life outside of Earth, and, thanks to Wells' superb novel, it was widely believed that alien life was probably hostile.

The arguments about the canals of Mars persisted almost until the photographs started to arrive from the *Viking* landers on Mars in 1976. Clearly there are surface features on the planet which in an Earth-based telescope give an illusion of linear markings, but this is only because the eye tends to interpret faint detail in a simplified form. If you draw a naked-eye view of the Moon, you may discover canals there as well, especially if you are told to look for them.

Although no Martians came running up to *Viking* to see what had invaded their planet, Wells' comments still contain some interesting suggestions. He implies that Mars is older than Earth (a few paragraphs earlier he says that Mars must have formed first out of the solar nebula because it is farther from the Sun). This may be right for the same reasons as it is now thought that the Moon rocks are older than those on Earth: the planet is much smaller than Earth.

Mars also gives the impression of a dying planet, although whether it ever supported life, even of a humble sort, is now very doubtful. The biological experiments carried out by the *Viking* landers indicated there were no living organisms where they landed, and although that does not preclude sparse pockets of simple life-forms such as bacteria surviving in special conditions, the possibility now seems so remote as to justify saying that there is no life on Mars. But Mars shows little sign now of the activity which once characterised it. The discoveries about Mars' past, and the final solution of the canal mystery, had to await the arrival of the *Mariner* spacecraft and the *Vikings*. Until then telescopic observation of the planet had revealed very little by comparison.

To the naked eye Mars has a distinct red colour. The planet is largely covered with reddish-brown rocks and dust which even colours the Martian sky pink. It is a *superior planet*, that is, farther from the Sun than the Earth. Therefore, according to Kepler's laws, it travels more slowly round its orbit than the Earth. When it reaches opposition, its closest approach to Earth, Mars' disc subtends an angle of about 25 seconds of arc, less than one seventieth of the Moon's diameter, and about half the apparent diameter of the lunar crater Copernicus. But that is only at its closest approach. Mars' orbit is rather eccentric, compared with most planets, as can be seen from the diagram on page 30. If Mars is at its closest to Earth (opposition) at the same time as its closest approach to the Sun (perihelion) it is almost twice as near to Earth as when opposition occurs when Mars is at aphelion (farthest from the Sun), and therefore its apparent size is also almost twice as great. At perihelic oppositions, which occur in August, Mars is 25″ in apparent diameter, compared with only 14″ at aphelic oppositions, which occur in February. This also means that at its closest to Earth, Mars can be well observed only from the southern hemisphere of Earth, as it is very low in northern skies. Mars keeps plodding along behind the Earth after opposition, easily visible to the naked eye, but maddeningly small in a telescope. Finally it goes behind the Sun at conjunction, and then the Earth has to continue for another year before it catches Mars once more.

The combination of the 2-year 50-day mean synodic period and the eccentricity of the planet means that observers on Earth have to wait 15 years between favourable oppositions of Mars. Then, when the oppositions occur near perihelion, Mars has another dirty trick to play: the planet often becomes shrouded in an enormous dust storm.

This happened during the last perihelic opposition of Mars in 1971 when *Mariner 9*, the first space probe to be put into orbit round another planet, was approaching Mars. Observers on Earth saw the planet growing larger in their telescopes, fascinating detail becoming clearer each day until, within a few days, the planet became featureless.

Mariner 9 sent back pictures of the planet as it approached, with the same frustrating result. The entire planet was covered in an enormous cloud of red dust. Why such storms apparently occur more frequently at perihelion can only be surmised. Clearly Mars gets a little warmer at such times, and this could trigger an atmospheric disturbance. Wind velocities of over 150km/h are needed to lift the dust, and it is thought that wind systems similar to the dust devils of terrestrial deserts may occur. Once the heated dust is lifted into the circulating air in large enough quantities, the sharp temperature contrast between the hot dust and the cold surrounding atmosphere sets up high winds at the edge of the storm, lifting more hot dust to feed the storm, until the entire atmosphere is full of dust. This would allow temperature differences to fall, winds to subside and, over a longer period, the dust to settle once more.

The effects of the Martian dust storms are to be seen in the pictures taken from the *Mariners* which often show dust patterns on the surface in the lee of surface features.

The great storm of 1971 did have two indirect benefits, however, in that the cameras on *Mariner* had little to do while the dust persisted, so they were turned on the moons of Mars, Phobos and Deimos. The second bonus was the gradual appearance through the dust cloud of a curious group of dark blobs. It was quickly realised that they represented very high mountains. Four of these peaks were aligned in a mountain chain known as the Tharsis Ridge, while a little to the west of this was the largest peak of all, a feature which had been named rather prophetically from Earth-based observations Nix Olympica, and which needed only a slight change of name to become the most appropriate Olympus Mons.

The unveiling of Mars

When the dust cleared, Olympus Mons was seen properly for the first time and caused no little excitement. This huge formation was clearly a volcano which had been active in the past, indeed might still be active as far as could be seen then. But it was far larger than any terrestrial volcano, some 600km across its base and towering 25km above the mean level of the surroundings. The other peaks in the Tharsis Ridge turned out to be volcanoes also. The entire area was in fact quite different from the cratered areas over which the fly-by *Mariners 4, 6* and *7* had passed, which had given the impression that Mars was rather similar to the Moon in appearance.

As further pictures were taken from orbit, it became clear that, while a high proportion of the southern hemisphere was heavily cratered like the Moon and Mercury, the northern hemisphere had extensive plains with very few craters, as well as the volcanic regions already mentioned.

Another startling feature was the great canyons and meandering, dried-up river beds suggesting that an extensive river system (perfectly natural, of course) had once existed. The polar ice caps of Mars, clearly visible from Earth in even small telescopes, are composed of water, and there may still be water permanently frozen beneath the surface of the planet.

When the two *Viking* spacecraft landed on the surface of Mars in 1976 the tentative conclusions being reached as a result of the *Mariner* observations were not much upset. It was confirmed that Mars is marked largely by volcanic activity in its remote past, with some extensive cratering remaining in some regions (although to a lesser degree than the Moon and Mercury).

The number of small craters remaining in good condition suggests that wind erosion is less than might be expected from the high-speed, dust-laden winds which occur so often. The *Vikings* measured the occurrence of isotopes of different gases in the Martian atmosphere, and the relative abundance of various isotopes suggested that Mars once had an atmosphere much richer in nitrogen than the Earth, and several other interesting differences in its early composition. These measurements clearly indicated the Martian atmosphere was once substantially denser than it is now. But Mars is a small planet, 6787km in diameter (about half the size of the Earth), and the force of gravity at its surface is only 38 per cent that of the Earth. So the gases in Mars' atmosphere have slowly been swept into space by the solar wind, leaving behind the present thin atmosphere consisting largely of carbon dioxide with traces of nitrogen, oxygen, noble gases and a little water vapour. There is enough atmosphere to form clouds, as well as to stir up the tremendous dust storms, but with a surface pressure of only 7mbar (compared with 1000mbar on Earth) and a daytime temperature of about 50°C below zero (and that only 25° north of the Martian equator) Mars is not a friendly place for Man.

Fear and Panic

Mars was known from Earth-based observations to have two small satellites. They were discovered by Asaph Hall of the US Naval Observatory 100 years ago, in August 1877. Named Phobos (fear) and Deimos (panic), the Greek names for the horses that pulled the chariot of the god of war, they became the subject of a well-known mystery—how did Voltaire, in *Micromegas* and Swift, in *Gulliver's Travels* know that Mars had two moons? Swift even had a pretty good value for their sidereal periods, and this was in 1726.

The 'obvious' explanation according to some authors is that he was told, by men from outer space. In fact, it is probable that Voltaire and Swift had heard of Johannes Kepler's guess that Mars had two satellites. Kepler argued that Venus had none, Earth one, Mars an unknown number, and Jupiter had four (the bright satellites seen by Galileo). As we have seen, Kepler still believed in a mathematical pattern to the Solar System, so it was logical to suppose that Mars would have two satellites (and presumably, also wrongly, that Saturn would have eight).

Little was known about the satellites, even up to the time of *Mariner 9*, because of their proximity to the planet. Phobos is so close, its mean distance from the surface being only 6000km (less than the diameter of Mars) and it is so faint (magnitude 11.6) that it is all but lost in the glare of the planet itself when observed from Earth. At this height above Mars, Phobos has to orbit the planet in only 7^{h} 39^{m}, so that it overtakes the planet's rotation on its axis. From Mars, Phobos would appear to rise in the west and set in the east, crossing the sky in an average time of $8\frac{1}{2}$ hours. The Martian day is only 40 minutes longer than the Earth's day, and its axial inclination is just over 1° greater, so the day-to-day behaviour of the sky would be very similar observed from Mars as from Earth, but Phobos would cross the sky the 'wrong way' during a single night's observation and would be a fascinating object in a Martian observer's telescope.

When *Mariner 9* had little to do while waiting for the dust storm of 1971 to subside, its cameras were turned on Phobos and Deimos. Phobos was found to be not even approximately spherical. It measures about 20 × 23 × 28km, and is something like a badly battered potato in appearance. The craters on the surface are of comparable size with craters found on other, larger bodies. Phobos has one, called Stickney (the maiden name of Asaph Hall's wife, whose persistent encouragement prevented her husband from abandoning what had seemed a futile search for the satellites), measuring 10km across, about half the diameter of the satellite itself. One of the most curious features revealed by the *Mariner* pictures was the presence of parallel grooves across one part of the surface. These have yet to be explained, although they clearly resemble the scars of a bombardment tangentially across the surface.

Phobos lies close to the *Roche limit* of Mars. This is the distance at which tidal forces of the primary body would prevent debris from forming a satellite body under mutual gravitational attraction of the particles, and which would cause a ready-formed body to begin to break up. The life of Phobos has been calculated as some 100 million years, after which it will finally crash down on to the surface of Mars.

Deimos is of a similar shape to Phobos, but even smaller, measuring some 15 × 12 × 11km. Its sidereal period is about $1\frac{1}{4}$ days, so that it orbits the planet in scarcely more time than the planet takes to rotate on its axis, with the result that Deimos is almost fixed in position in the Martian sky while the stars carry on their daily rotation behind it.

As expected, both satellites keep one face turned towards Mars, just as the Moon keeps one face turned towards the Earth. Both appear to be of grey material quite different from the red rocks of Mars. The speculation on their origins continues, since while they appear to have all the expected characteristics of asteroids, irregular shape and heavy marking from collisions, they are in almost circular orbits in the plane of Mars' orbit, suggesting they are the remains of the 'rings of Mars', that is, the rings of debris which probably surrounded all the planets during their formation.

The missing planet

The legacy of the Greek philosophy of a mathematical explanation of the Universe persisted right through the Renaissance. Kepler found that if a sequence of spheres were postulated each enclosing the regular solid geometrical forms (octahedron, icosahedron, decahedron, tetrahedron and cube) the orbits of the planets fitted around these spheres. But a more familiar sequence was noted by Professor Johann Daniell Titius (1729–1796) and also by Christian Wolff (1679–1754). One takes the mathematical sequence 0,3,6,12, etc. and adds 4 to each, thus: 4,7,10,16,28 etc. The series represents the relative mean distances of the planets from the Sun, the distance Earth to Sun being 10. If the series is divided by 10 it gives the approximate values in astronomical units. This 'law' was enunciated by Johann Bode (1747–1826) and is referred to by his name.

Bode's Law has no defined basis; it is merely a mathematical curiosity. It successfully predicted the mean distance of the next planet to be discovered, Uranus, but failed completely with the next, Neptune. There was also a startling gap between Mars and Jupiter. Not only did Bode's Law suggest that there should be a planet at a mean distance of 2.8 astronomical units, but the very size of the gap is startling, as can be seen from the scale diagrams of the Solar System, page 30.

Despite a systematic search by a number of astronomers in different countries towards the end of the eighteenth century, it was not until the first day of the new century that Fr Giuseppe Piazzi discovered a peculiar 'comet' which, he noted, moved from night to night among the stars. He noted its position carefully, but by the time other astronomers tried to find it, it was approaching conjunction, and too close to the Sun. But from Piazzi's observations the mathematician Gauss was able to calculate its position in the sky for the following year, when in due course it was found again. It was about the right distance to fill the missing position in the Bode series, but it was scarcely a planet to get excited about, being only of magnitude 7.5 at its brightest, invisible to the naked eye. The new planet was named Ceres. After Ceres' discovery, astronomers were still unconvinced that this was the missing planet, and the search continued. Just over a year later, another tiny planet was discovered, Pallas, and within two years another, Juno, was found; thereafter to the present day the asteroids have been discovered in their thousands.

Ceres has since been found to be only 955km in diameter, and the next largest, Pallas, is only about half this size. But asteroids have now been found with orbits of considerably different eccentricities and inclinations to the plane of the Solar System, suggesting that many of them are debris left over from the formation of the Solar System. But by far the largest number are concentrated in the Mars–Jupiter gap, and the mean distance of all these is just about the 2.8 AU predicted by Bode's Law.

Examination by spectrographic techniques and comparison of the results with the composition of meteorites suggests that about a tenth of all asteroids resemble stony iron meteorites. These are generally fairly large and are found in the inner parts of the 'asteroid belt' between Mars and Jupiter. Carbonaceous chondritic material, of the type found in many meteorites which contain the types of material of which the Solar System was formed, is detected in almost all the remaining asteroids.

It is thought that a small planet may have been gradually forming at a late stage in the evolution of the Solar System. The gravitational effect of Jupiter, which trapped some asteroids in groups both leading and trailing the great planet in its own orbit, and which has also caused gaps to be formed in the asteroid belt (much as there are gaps in Saturn's rings), may have caused too much disruption. Alternatively the collisions

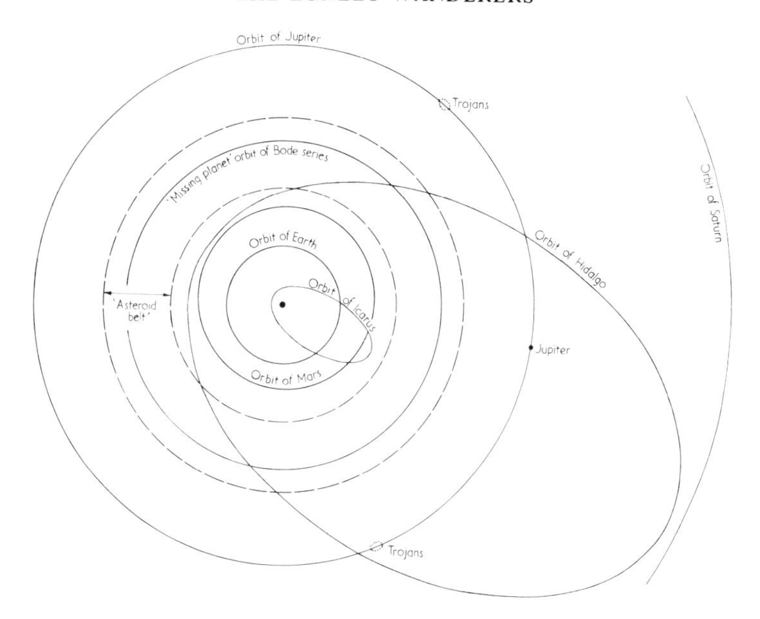

14 Most of the asteroids are in a zone between Mars' and Jupiter's orbits, the 'Asteroid Belt', near the centre of which is the missing distance in the Bode series. Examples of other asteroids, such as Sungrazer Icarus and the Trojans which share Jupiter's orbit, are shown.

between the small bodies may have simply caused the protoplanets to break up into smaller lumps. Certainly there is plenty of evidence to suggest that the irregularity in shape of Mars' satellites may be nothing compared with the shapelessness of some of the smaller asteroids. This can be seen by the considerable variation in the brightness of many of the asteroids as they turn different faces towards Earth. As well as variations in the amount of scarring on the surface, which could have a fairly small effect on reflectivity, the irregularly-shaped asteroids will actually appear to change in size.

The irregularity of asteroids and small satellites makes estimating their size very difficult, since this has to be based on measurements of brightness and an assumed average reflective index for the surface of the body. Thus, say, a sausage-shaped asteroid or satellite will be much dimmer when end-on to the observer than when side-on.

A stir was caused in autumn 1977 when the press proclaimed the discovery of the Solar System's tenth planet. This was a very faint object (magnitude 18, that is over 63 000 times dimmer than Uranus, which is only just visible to the naked eye) found on a photograph of the sky in the region of the constellation Aries by the Palomar astronomer Charles Kowal.

It was found that the new 'planet' was between the orbits of Saturn and Uranus, and although it is too small to be properly called a planet, its estimated diameter is 300 to 400km, about the same size as Ceres or Pallas, the largest of the asteroids. There are only about 20 asteroids with diameters of 200km or more, and these are all found in the asteroid belt between Mars and Jupiter, at about the Bode position of 2.8 AU, so clearly the tiny body discovered by Kowal is very unusual. There are a number of places in between the planets, and even on their orbits, where stable orbits can be found for other bodies. The best-known example is the two groups of asteroids trapped in Jupiter's orbit 60° ahead and behind the planet itself, called the Trojans. Some of the asteroids have such eccentric orbits that they are taken from close to Mars' orbit to well beyond Jupiters, the most famous such asteroid being called Hidalgo. The discovery by Kowal probably means that there are several other small bodies in various distant parts of the Solar System.

The star that never was?

The Solar System contains two broad types of planet: the rocky terrestrial type, which are small, closer to the Sun, and of high density; and the gaseous giants, which are very large, of low density, and more distant from the Sun. Of these outer planets, Jupiter is by far the biggest and most important. At the time of writing, Jupiter is the most distant planet that has been visited by a space probe, and in many ways the pictures sent back were less of a surprise than any sent from the inner planets. Jupiter close up looked very much as might be expected from examination in Earth-based telescopes. Dense, brownish clouds were seen, covered in layers of lighter gases, and formed into belts around the planet by its enormous rate of rotation. Here and there reddish 'whirlpools' of gas form the distinctive spots, including the Great Red Spot itself.

However, the pictures returned by *Pioneer 10* and *11* were most dramatic, and the amount of information received from these probes exceeded all the previous knowledge of the planet.

Jupiter has a mass 318 times that of the Earth, which is not very much when compared with its volume, over 1300 times that of the Earth. This is because Jupiter is composed mainly of the lightest elements, hydrogen and helium. It is the stuff stars are made of.

We have already seen that Jupiter contains nearly $2\frac{1}{2}$ times the mass of the rest of the planets put together. There is good reason to suppose that this planet, and those like it, are composed of the original material of the Solar System. Perhaps, if Jupiter had been even larger, its enormous mass would have been sufficient to cause it to collapse under gravitational forces and so to generate enough energy to become a star in its own right. The Sun would then have been a binary star, like so many others throughout the Galaxy.

Hydrogen, helium and the other elements and compounds detected on Jupiter, such as ammonia, methane and water, are basically colourless, and other compounds which might be present, such as ammonia compounds, organic compounds and other complex molecules, would quickly be swept into the higher-temperature regions of Jupiter's thick atmosphere by the continually raging storms in which these more complex substances would be broken down into simpler ones. The mechanism remains unknown at present, but has given rise to the interesting suggestion that organic compounds, and even primitive life, may exist in Jupiter's atmosphere.

Because of its low density, the planet may be regarded as nothing but atmosphere. But the *Pioneers* have shown that Jupiter has an intense magnetic field; this had already been supposed from the fact that the planet emits strong radiation at radio wavelengths. Planetary magnetic fields are thought to be associated with some form of magnetic core acting like a dynamo owing to the planet's rotation. Jupiter would certainly be efficient in this respect, since it rotates once in about 10 hours. This means that the atmosphere of Jupiter is travelling at some 44 700km/h at the equator.

Despite this tremendous speed, Jupiter's gravitational field binds even the lightest element, hydrogen, to the atmosphere, whereas in all the inner planets with any atmosphere, the lighter gases escape into space and are swept away by the solar wind. But this rapid rotation causes the planet to bulge very noticeably around the equator. At higher latitudes the atmosphere moves more slowly, so that Jupiter's rotation period can be judged by features at the equator to be 9^h 50^m 30^s, or at higher latitudes to be 9^h 55^m 40^s. Radio emissions from Jupiter's interior are polarised, and the polarisation varies with Jupiter's rotation in good agreement with the higher-latitude figure. This would suggest that the interior of Jupiter rotates once in about 9^h 55.5^m, but leaves it a mystery why the atmosphere at the equator rotates faster than the interior of the planet beneath.

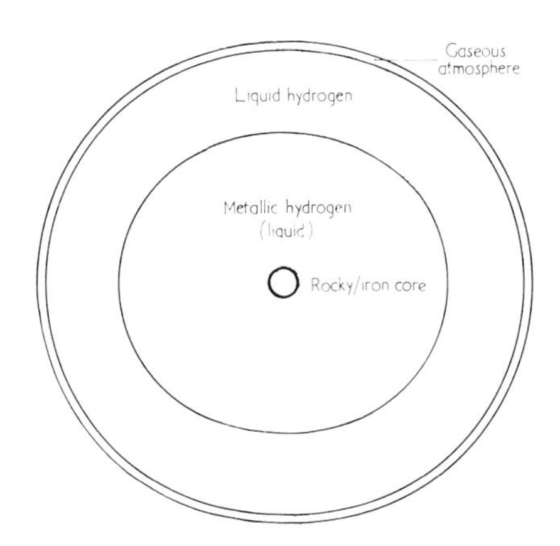

15 A cross section of Jupiter, showing that the expected composition of the planet is mostly hydrogen.

Jupiter's gravitational field produces enormous pressures beneath its surface atmosphere, and it is thought that most of the interior is probably liquid hydrogen with only a small core of heavier materials. The depth of the liquid-hydrogen layer could be some 24 000km, causing such colossal pressures that at the bottom of this layer the hydrogen would behave like a metal. This metallic liquid-hydrogen core would be a good electrical conductor and could produce the magnetic and electrical fields observed.

Jupiter has a large retinue of satellites. At least 13 are known, namely (from Jupiter outwards) Amalthea, *Io, Europa, Ganymede, Callisto,* Leda, Himalia, Lysithia, Elara, Ananke, Carme, Pasiphae and Sinope. Only four of these satellites (those in italics) have been long known. These were the four discovered by Galileo when he first turned his telescope on Jupiter. Not only was it clear that Jupiter had four satellites, but it was obvious that they revolved around Jupiter, not around the Earth. This was the first direct observational evidence that the Earth was not the centre of all things in the Universe.

Two of the largest satellites, Ganymede and Callisto, are larger than Mercury (their diameters being estimated at 5270 and 5000km compared with Mercury's 4880km). Although Jupiter's satellites are of planetary size, they are different in many important respects from the inner planets, and from each other. Ganymede and Callisto are both of low mean densities, 2 and 1.6 times the density of water respectively, while Io and Europa are about 3.5, similar to the Moon and Mars. This is due, it is thought, to thick layers of water frozen into the crust of the satellites with varying amounts of dense rock in the ice layer. This would mean that Io is almost entirely rock, while Ganymede must have a solid core about the same size as Io.

Io is a fascinating satellite in itself. Its sides have been shown to be of unequal reflectivity, and it is distinctly red. It also has red polar caps. While Io has a high reflectivity (*albedo*), its spectrum does not indicate infrared absorption—another reason to suppose that there is little ice on the surface. The fact that the ice content of the satellites increases with the distance from Jupiter suggests that in the early history of the system Jupiter produced a great deal more heat than it does now, thereby driving off all but the rocks of Io, most of the water from Europa, less from Ganymede, and so on. (Little is known about Amalthea, which is a relatively recently discovered body only some 150km in diameter.) The list of possible substances that could provide Io with its observed albedo is limited. Various salts are candidates, as they could have been deposited on the surface by water which was subsequently driven off by the heat from Jupiter, leaving a salt-encrusted surface.

Io has for some time been known to be associated with periodic variations in the radio signals emitted by Jupiter itself, the variations apparently depending upon the position of Io in its orbit. The simplest explanation is that Io has an ionosphere, a surrounding layer of gases highly reflective to radio waves.

In 1973, Robert A. Brown of Harvard University discovered sodium in the emission spectrum of Io. It was found that the sodium extends far into space, and is present around the entire orbit of the satellite. This suggests that Io itself is pouring sodium into space. If Io is a rich source of sodium, it may well be that the salt covering the surface is common salt, sodium chloride.

When *Pioneer 10* approached Jupiter, it was found that there is an even larger cloud of hydrogen surrounding the orbit of Io. Since then, other elements have been detected in this region, including potassium and ionised sulphur. The conclusion appears to be that Io has a very tenuous atmosphere of ionised gases, driven from the surface of the satellite by the powerful radiation of the particles trapped in Jupiter's intense magnetic field.

The four innermost satellites (excluding Amalthea) are all bright enough to be seen through a modest pair of binoculars. They all lie close to the ecliptic, so that they appear to be more or less in line from Earth, and all but Callisto must pass into

Jupiter's shadow and cross the disc of the planet each revolution. They are fascinating to watch by means of even a small telescope. Jupiter is not only the king of the planets, but of all the planets provides the most colourful and varied views.

As far as the eye could see

While Jupiter may hold the observer's interest for longest with its wide variety of features, the planet which for so long marked the outer known limit of the Solar System, Saturn, must rank among the finest sights in a telescope. Before the invention of the telescope, Saturn was as far as the eye could see in the Solar System. (In fact, Uranus is just visible to the naked eye if you know where to look, but it was never spotted until the telescope had first picked it up.)

Saturn is a bright object in the night sky, yet Uranus, with just under half Saturn's diameter, is almost invisible. This well illustrates how the planetary distances from the Sun increase dramatically. Bode's series becomes approximately a doubling of the previous distance after Jupiter.

Saturn has two distinct claims to fame: its rings, the most familiar telescopic objects in astronomy, and its density, the lowest of all the planets.

The planet itself is 119 300km diameter at the equator, 84 per cent of the equatorial diameter of Jupiter, so it is comparable in size as Jupiter's neighbour. Yet Saturn is only 95 times as massive as the Earth, while its volume is 755 times as much. If we could find a lake big enough, Saturn would float high in the water, for its density is only 70 per cent that of water.

Clearly the composition of Saturn must be mostly hydrogen, and any rocky core which may have been the nucleus of the planet as it was being formed is unlikely to be larger than 20 000km, only about 50 per cent larger than the Earth. It is suggested that this core is covered mostly with liquid molecular hydrogen, although there may be a fairly thin layer of metallic hydrogen and even of ice around the core. Earth-based observations of Saturn show bands around the planet similar to those on Jupiter, but only a pale imitation, possibly because Saturn's atmosphere contains fewer compounds of ammonia. Ammonia is not detectable in the atmospheres of Uranus and Neptune, presumably because on these planets the temperature is far too low for the compound to be present in the form of a gas.

Saturn rotates on its axis once in 10^h 14^m, so it also bulges noticeably at the equator, but, unlike Jupiter, it is inclined to the plane of its orbit by a fairly large angle, $26\frac{3}{4}°$ compared with Jupiter's 3°. This is fortunate for observers on Earth because on most occasions it results in an impressive display of the rings. However, the satellites of Saturn are much less easily seen than Jupiter's, unless the Earth happens to be passing through the plane of the rings, when the rings themselves disappear and the satellites are aligned (although they are still very much fainter than the Galilean satellites of Jupiter).

The temporary disappearance of the rings in our telescopes occurs because the rings are so thin compared with their diameters. Estimates have put the thickness of the rings at only 15km and when presented edge-on this is quite impossible to detect at the distance of Earth, except for the shadow of the rings on the disc of Saturn itself.

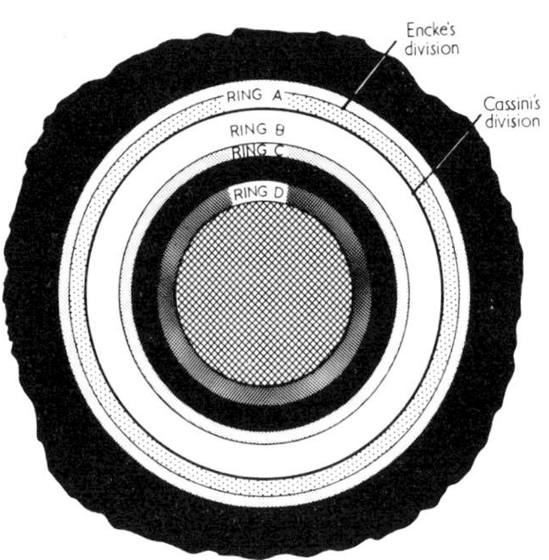

16 The rings of Saturn as never seen from Earth (compare the photograph on page 85) looking down on one of the planet's poles.

The composition of the rings is still unknown, although it is clear that they are formed of thousands of tiny particles, and spectroscopic investigation has indicated that these may be ice particles, or at least ice-covered. But whether the rings are the remains of a satellite that was unable to form into a solid body, or the result of a satellite breaking up, or colliding with another body, still cannot be resolved. The gravitational forces of the rings themselves and of Saturn have caused the ring-system to divide into at least five clearly defined rings; ring A, the outermost, which is divided into two by the so-called Encke division; ring B, separated from A by the larger Cassini division, and brighter than A; ring C, the so-called crêpe ring, very transparent and shadowy; and the innermost ring D, which almost clings to the surface of Saturn and is very faint indeed.

Saturn has ten named satellites: Janus, Mimas, Enceladus, Tethys, Dione, Rhea, Titan, Hyperion, Iapetus, and Phoebe. Of these, only Titan is comparable to the major Jovian satellites, and indeed may well be the largest satellite in the Solar System at 5800-km diameter. Titan is larger than Mars, and possesses an atmosphere of methane and hydrogen with a surface pressure of as much as one tenth that of the Earth. However, its density is only 30 per cent greater than water, so at a temperature of –150°C probably contains a great deal of ice.

The next to outermost, Iapetus, is next largest of Saturn's satellites at 1600-km diameter, and its orbit is fairly steeply inclined to the plane of the planet's equator (and the rings). This satellite has caused much interest because the albedo of one side appears to be six times higher than the other. Such an asymmetrical appearance suggests either a fantastic collision on one side, or, as has been suggested more recently, by a continuous bombardment on the side away from Saturn (it is assumed to have captured rotation) from small particles still spiralling inwards towards the planet from the debris associated with the outer satellite, Phoebe, which orbits Saturn the 'wrong' way (most bodies orbit their primaries anticlockwise, as seen when looking down on the north pole of the primary) so that any such debris could strike Iapetus at very high speeds.

Herschel's planet

Sir William Herschel (1738–1822) discovered Uranus in 1781 without realising at the time that it was a new planet. For thousands of years it had been thought that Saturn marked the outer boundary of the Solar System and though Uranus had been spotted many times before, no one had detected its movement relative to the starry background which singled it out as an object of the Solar System. Herschel assumed it was a comet, as its appearance in the telescope clearly showed it was not a star. When the orbit was computed, and it was found not only to be a new planet, but to be just where the Bode series suggested it should be, it caused a sensation.

Not much detail can be seen on Uranus: seen in Earth-based telescopes it is virtually featureless. Its density suggests that it is also a small rocky-cored planet deeply covered with liquid hydrogen. It is too small, about 50 000-km diameter, to compress the hydrogen to a metallic state, and it probably contains a thick layer of ice around its core. Its five known satellites are all rather small: Miranda 500km, Ariel 1500km, Umbriel 1000km, Titania 1800km, and Oberon 1600km (the sizes of Uranus' satellites are very speculative) but all these satellites orbit very precisely in the plane of Uranus' equator, fairly close to their primary.

Because of its uncertain size, the opportunity to observe a transit of the planet across a star, a very rare event predicted for a night in March 1977, caused a great deal of interest among astronomers. The event when it occurred resulted in a fascinating discovery; the planet was approaching the position of the star when the star dimmed—some 10 minutes before the planet's disc had reached the star. Before the occultation finally took place, the star dimmed sharply a total of five times. This pattern was repeated symmetrically as Uranus moved away from the star. The explanation is clearly that Uranus has rings too!

Plate 10: Saturn and its ring system photographed with the 100-inch Mount Wilson telescope. *(Photograph from the Hale Observatories)*

Plate 11: Typical stellar spectra for the cooler types of stars compared with the Sun's spectrum (the star class is given on the left of each spectrum, and the star name or number on the right). *(Photograph: Department of Astronomy, University of Michigan, Ann Arbor, Michigan)*

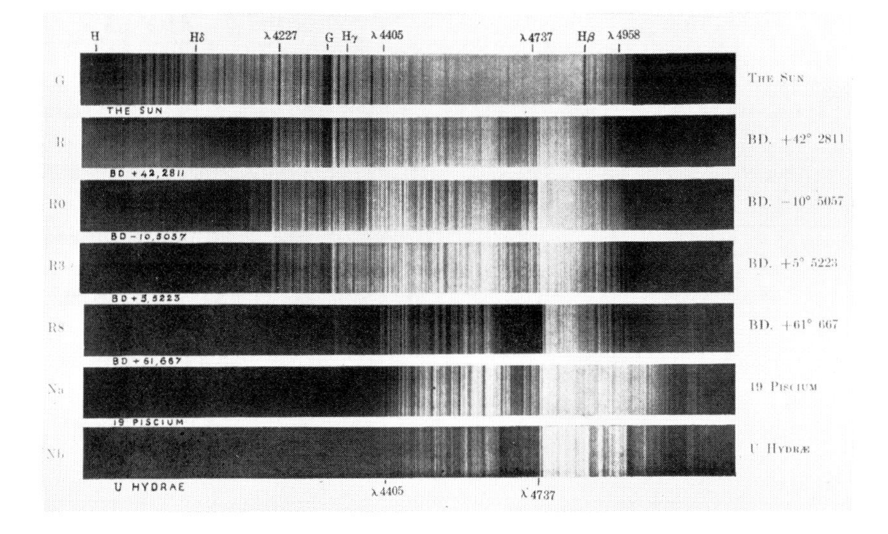

Plate 12: The Great Nebula in Orion, M42 (NGC 1976), and its satellite nebula. This emission nebula contains many young stars. The dark lanes of dust and gas may be the stuff the stars are made of. A number of small dark globules (a few light years across) can be seen. *(Photograph: Lick Observatory)*

Plate 13: The Crab Nebula in Taurus is the remnant of a supernova observed in 1054 AD. This photograph in red light, taken with the Hale 200-inch telescope at Mount Palomar, shows the incredible structure of this rapidly expanding cloud of glowing gas, powered from a pulsar in its centre.

At the time of this occultation, the rings of Uranus, assuming they are in the plane of its equator like the satellites, would have been fairly 'open', approximately 50° inclined to the Earth. This was lucky because had the rings been closed we might have had to wait a very long time before another suitable opportunity arose.

The problem is that Uranus is inclined to its orbit by over 90°. In about 1986 it will present its north pole to the Sun, so that if you could stand on Uranus' north pole, the Sun would be a brilliant star (actually a disc only 1.7 minutes of arc diameter) overhead. The Sun would not set until 2008, and then you would not see it again until 2050! The satellites (and the rings, although they are too faint to detect normally) appear from Earth to go round the planet in circles at some dates (1946, for example, and again, but in the opposite direction, in 1986), then in ellipses, getting narrower until the satellites are in line (in 1966 and 2008). They also move north and south of the

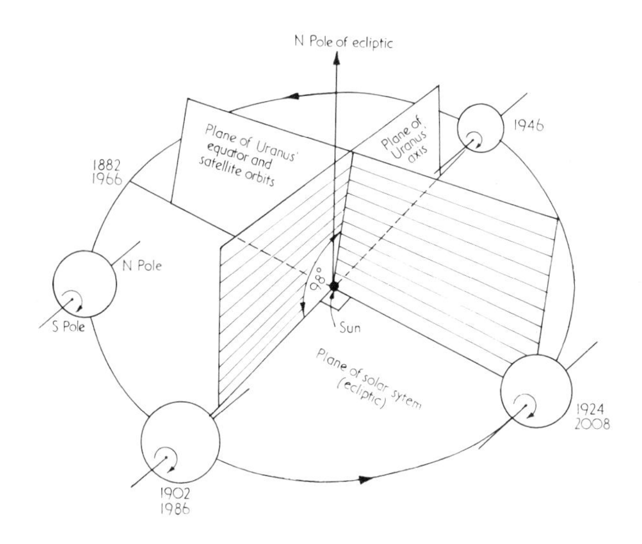

17 Uranus' poles are inclined at 98° to the ecliptic so that the planet presents first one pole to the Sun (and the Earth), then its equator, then the other pole, and so on, and its satellites appear to reverse their direction of rotation about the planet.

planet as seen from Earth, rather than east and west as most satellites appear.

The rings are unlike Saturn's, appearing to be in five or six narrow belts, possibly locked in orbits by the gravitational action of the satellites themselves. Why does Uranus have rings, and why is it at such an unusual angle? Apart from Venus, whose 'north' pole points in the general direction of the south pole of the ecliptic, Uranus seems the odd man out in this respect. It has been suggested that early in its history Uranus was in collision with a fairly large body that knocked the planet on its side. If so, perhaps the rings are the remains of the culprit.

The misplaced planet?

Bode's Law scored its final success with Uranus. It was not long after Uranus was discovered that it became clear that the new planet was not following its predicted path accurately enough, and that something was disturbing its calculated orbit. The story of how the search for another new planet began, and the exciting climax of this search is worth reading at greater length than space permits here.

The position of the supposed new planet was calculated by John Couch Adams in 1845, who, having no telescope of his own, sent his calculations to the Astronomer Royal, George Airy. Airy ignored the student's calculations. The following year, a similar set of calculations were published by the French mathematician, Le Verrier. This prompted Airy to start searching the area predicted by Adams the year before, and at his request, charts of the area were drawn up by two astronomers, including one James Challis of the University of Cambridge. Challis' observations in fact included the new planet, but he failed to spot the intruder on his charts. Meanwhile, the Berlin observatory was working with Le Verrier's figures and in September 1846 Galle and d'Arrest became the official discoverers of Neptune. There are a good many lessons in that story!

The discovery was a triumph for theoretical astronomy, although it is interesting to speculate on how the story might have been different had Neptune been on the opposite side of the Sun to Uranus at the time of Herschel's discovery of the latter, since the presence of Neptune would then have been unpredictable. As it was, Bode's Law had failed miserably for the first time. Neptune should be 38.8 AU from the Sun according to Bode's Law, but it turned up at 'only' 30. There were also some unexplained variations in Neptune's predicted orbit, so the search began all over again. This time, it was never to be successful. Although Pluto was found earlier this century at a mean distance of just less than 40 AU, it has such an eccentric orbit that the apparent reasonable fit with Bode's Law can be discounted.

Neptune is very similar in size, and probably in composition, to Uranus. But while Uranus may be picked up easily in binoculars, and a moderate amateur telescope shows Uranus' tiny, wan disc, Neptune only just exceeds magnitude 8 at opposition, and appears star-like in all but fairly large telescopes.

Two satellites have been found. One of these, Triton, is possibly as large as 6000km diameter (larger than Titan) and orbits the primary in a retrograde direction. Pluto is about 5800km diameter, and its orbit is so eccentric that, at perihelion, Pluto is closer to the Sun than Neptune, and it is often suggested that Pluto may once have been a satellite of Neptune. Models have been worked out in which Pluto could have broken free in a gravitational dust-up with Triton, which left it orbiting Neptune the wrong way round.

Neptune's other known satellite, Nereid, is estimated to be only 500km in diameter.

The planet at the frontier

Pluto was discovered, as many important astronomical discoveries have been made, by accident. At the beginning of the twentieth century, Percival Lowell at Flagstaff examined the orbits of the outer planets, as Adams and Le Verrier had done with the orbit of Uranus, to try to find the cause of still unexplained perturbations in the orbit of Neptune. Using similar mathematical procedures to those of Adams and Le Verrier, Lowell predicted that another planet would be found beyond Neptune.

Lowell himself died before the search ended, and the predicted planet remained a mystery for another 13 years until in 1929 the staff at Flagstaff began the search again. The following year the search bore fruit. A tiny mark on a photograph was spotted by Clyde Tombaugh. This tiny planet could not explain the perturbations that Lowell had used as the basis for his calculations, and this has given rise to the question of whether there might be yet another planet sufficiently massive to betray its presence by its gravitational effects, yet still invisible. Pluto's orbit is highly inclined to the ecliptic and it might have been very much harder to find had it not coincidentally been close to its ascending node (the point in an orbit where the body passes from south to north of the ecliptic). The orbit of a tenth planet might be even more inclined to the ecliptic and hence it would be very difficult to track down. But if such a planet is of significant size, it is to be expected that by now, after half a century or more of extensive celestial photography, it would have been picked up. For the moment, then, Pluto remains the outermost planet.

Pluto's diameter is estimated to be about 6000km, smaller than Mars. Its distance from the Sun ranges from 4425 million kilometres to 7375 million kilometres (a distance which light takes nearly seven hours to cover). It also has the dubious distinction of being the coldest planet, with an estimated surface temperature of about –230°C, and a year 247.7 times as long as Earth's.

Since its discovery nearly 50 years ago, it has moved across the sky from where it was discovered, in the middle of the constellation Gemini (the constellation in which Herschel discovered Uranus) into the far north of the constellation Virgo—about the same distance across the sky that the Moon covers in five days. It is already some 11° north of the ecliptic, and in the 1980s as it approaches perihelion (for the first time since it was discovered) it will also be almost its furthest north of the ecliptic, 17°. Although closer to the Sun than Neptune, it will also be a long way north of Neptune's orbit, some 1300 million kilometres (almost as much as Saturn's mean distance from the Sun). So there is little chance of a collision between Pluto and Neptune! In fact this enormous separation makes many astronomers dubious about the possibility that Pluto was once a satellite of Neptune.

Messengers of the stars?

In spite of its tremendous distance from the Sun, Pluto is by no means the outermost of the Sun's retinue. The real long-distance travellers are the comets. Some of these eventually make a turn of the Sun and head out of the Solar System, perhaps never to return.

Comets are often misunderstood by the non-astronomically minded. Spectacular photographs and vague stories about Halley's comet have led many people to imagine a comet is a brilliant object which streaks across the sky in a few seconds with a long train or tail flowing behind in its wake. This is a fair description of a fireball, which is an unusually large meteor entering the Earth's atmosphere, but is not like a comet at all, which is usually many millions of kilometres away, in orbit round the Sun. Most comets are very faint indeed, and if they are bright it is usually only when they are close to the Sun. This is because comets shine by reflected sunlight, and also radiate light when stimulated by the incident solar radiation. But, in either case, the comet needs to be relatively close to the Sun to shine with any brilliance. This usually presents the same sorts of limitation on naked-eye observation as apply to Mercury, Venus, or other objects close to the Sun in the sky.

Many proposals have been made to account for the composition of comets. Now, most astronomers agree that they are very tenuous bodies, probably containing a core of small dust-like particles. Professor F. L. Whipple has proposed that comets contain a great deal of water frozen into what he describes as a 'dirty snowball'. In many respects, the constituents of comets would resemble those of Uranus, Neptune and some of the larger outer planetary satellites, according to present models of these planets. They show similar spectroscopic results, have similar densities, and are almost all at the outer boundaries of the Solar System.

In fact, as so many comets visit the Sun each year (most of which are quite invisible to the naked eye) it follows that many, many more remain in the outer reaches of the Solar System, and it has been suggested that Uranus and Neptune were formed of the same basic materials as the comets while Saturn and Jupiter were composed largely of the original material of the Solar System, hydrogen plus some heavier elements bound in compounds such as ammonia.

The idea of a vast cloud of comets (meaning just their nuclei) at distances approaching a light year away from the Sun was proposed by J. H. Oort of the University of Leiden in the early 1950s. External influences, he said, such as a 'close' approach of another star (within a light year or two) could disturb some of the Oort cloud material causing it to fall towards the Sun, thereby acquiring the highly elliptical orbits most comets follow. Some of these comets would be influenced by the particular planetary configurations of the Solar System at the time of their close approaches to the Sun, causing some to move into closer, more stable orbits with comparatively short sidereal periods, while others would be pulled into orbits taking them out of the Solar System for ever. It follows, of course, that the space between the stars may be traversed by countless millions of comets, but only a very few would be likely to fall under the Sun's influence over many millions of years.

Comets usually fail to appear precisely on time, even those relatively close to the Sun, such as Halley's comet, which takes about 76 years to orbit the Sun, passing just outside the orbit of Neptune at aphelion. This is thought to be a result of the streams of ionised particles driven off by the solar wind which, combined with the rotation of the comet's nucleus, causes a small pressure to be exerted by this 'jet' either towards or away from the direction of motion, so sometimes speeding up, sometimes slowing down the comet. About as many comets arrive early as late, so some normally random factor is likely to be involved: in this case, the direction of rotation of the nucleus.

As the comet approaches the Sun, both gases and dust particles are 'blown' out of the nucleus. The ionised gas absorbs radiation from the Sun, which causes it to glow. This forms a tail pointing away from the Sun, so that after perihelion a comet travels tail first. The Sun emits large volumes of charged particles forming the solar wind. Owing to the rotation of the Sun itself, this 'wind' is emitted almost tangentially at the Sun's surface, moving outwards radially like the sparks thrown from a Catherine Wheel. This means that close to the Sun, the solar wind does not 'blow' radially, but as the distance from the Sun increases, the more truly radial the pressure of the solar wind becomes. This means that the dust particles in the comet are subjected to a pressure from a direction slightly different from that of the incident radiation forming the ionised tail, so that the comet often has its tail spread out in a fan shape, or split into two distinct tails, one full of glowing gas showing plenty of the 'structure' characteristic of radiating gases, and the other of more amorphous appearance formed by reflected light on the dust cloud.

As yet, no space probe has been sent into the tail or nucleus of a comet. We do not even know if it is possible to pass through the nucleus, or at least its outer regions. Certainly, the tails of comets are virtually nothing! The Earth has passed through them many times, but has it ever encountered a nucleus of a comet? It is possible that the famous Siberian meteorite of 1908 was a small comet nucleus, or perhaps a small asteroid—or perhaps this means the same thing. It is unlikely that the nucleus of a comet is very large: none has been observed in transit across the Sun's disc, for example, although this still allows comets to have plenty of mass.

It is known that meteorites are associated with comets, and as well as debris shed by the nucleus over the life of the comet, following the comet in its orbit, perhaps the ultimate fate of comets trapped by the Sun is to be stripped of all gases, water or other frozen compounds, and to become just another collection of tiny particles.

As in so many branches of astronomy, the problems of analysing the Solar System are problems of building better models to fit the very scanty facts at our command. It is seldom that even the best models do not have obvious contradictions. There are almost as many theories of the origins of the Solar System as there are astronomers, but a pattern is now emerging, linked with even more speculative thoughts about the origins of the stars and galaxies themselves. We will return to the questions of origins, and to the ultimate fates of the planets and stars, in later chapters. Let this chapter serve to illustrate the diversity and immensity of our own 'back yard' in space because, when we consider the stars, it will become clear that 'you ain't heard nothin' yet'.

3 The Depths of the Void

The Renaissance astronomers shed the long-held, comfortable vision of the unchanging, almost clockwork Universe of Ptolemy with great reluctance. The Church preached of a rather peculiar Creator who made a large celestial machine with which to astound his chosen creatures, who lived on a single, small, Earth at the centre of the machine. It is difficult to appreciate the intellectual upheaval necessary to challenge this vision of the Universe, and in the end to show that creation—and hence the Creator—was vaster, stranger, and more wonderful than Man had ever been able to imagine.

One of the notions which had to be dispelled was the unchanging nature of the stars. As the Earth speeds round the Sun, the stars appear to remain 'fixed' on the celestial sphere even when examined through a high-power telescope, and therefore, one can argue, they must all be the same distance away.

For the ancient astronomers, of course, the problem was simply that the distances to most of the stars were so much greater than expected that it was some time before astronomers realised that the parallax of even the nearest stars would be very small indeed—far too small to detect by eye alone even with a telescope.

The search for a star close enough to show a shift in its position on the celestial sphere due to parallax occupied many eminent observers, and led to several important incidental discoveries on the way. The Reverend James Bradley, who was to succeed Halley as Astronomer Royal, found in 1728 that a star he was measuring in the constellation Draco apparently moved, but not in the way expected from parallax. He deduced that the displacement was due to the motion of the Earth round the Sun and the effect this had on the relative position of the star due to the finite velocity of light. The effect, called *aberration*, displaces all celestial objects from their true positions by up to 20.4″ (one-third of a minute of arc) depending upon the direction the Earth happens to be moving, and is similar to the apparent shift of the wind direction in a fast-moving sailing boat.

Another technique was tried by Sir John Herschel, who argued that stars could be of very different distances, their proximity in the sky being due to line-of-sight effects. The parallax of one relative to the other should be easier to detect than parallax of an isolated star. His study of the positions of double stars led him to find that they were in fact gravitationally linked in many cases.

A third technique was tried by a German astronomer, Bessel, in 1838. By this time it was known that the celestial sphere was by no means immutable; stars varied in brightness (that had been known even by the ancients, but was explained away as trivial or magic), and new stars were born, sometimes of incredible brightness for a very short period. But most important of all, some stars were in motion across the celestial sphere. It was by now realised that stars like our own Sun are moving through space, and though this *proper motion*, the star's apparent change of position, was very small, it could be measured. Bessel argued that the nearest stars would show the greatest proper motion, and so he chose a faint star in the constellation Cygnus, the Swan, known to have an annual displacement of just over 5″. He found that the position of the star, taking its proper motion into account, varied by about 0.6″ when observed at intervals of six months. This meant that the star shifted relative to the background of the distant Milky

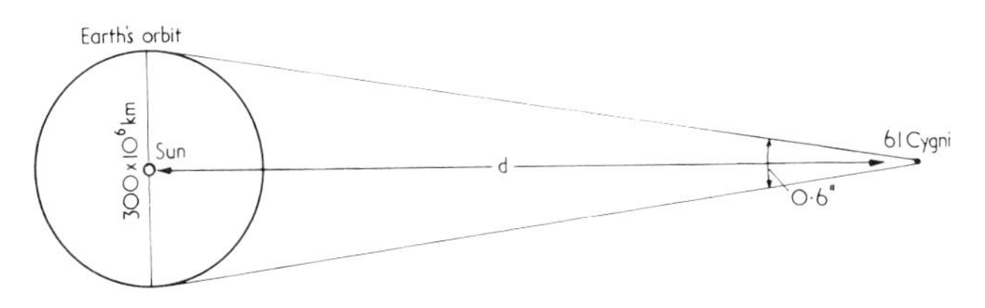

18 The principle of stellar parallax measurement. For 61 Cygni (see star chart 6) the angle is only 0.6″.

Way stars between the two observations, from two points separated by the diameter of the Earth's orbit, 300 million kilometres. This was the key astronomers had been searching for. The distance to 61 Cygni, Bessel's faint star, could now be calculated by solving the triangle of distance d to the star, the length of both sides, and of base B, 300 million kilometres, having an angle of 0.6″ at its apex. If the angle is turned into radians, the solution is simple, since the sines or tangents of very small angles are equal to the angle itself. To change degrees to radians we must multiply by $\frac{2\pi}{360}$ and to change seconds of arc to degrees we divide by 3600. The distance to the star is given by

$$d = \frac{B}{\tan p} = \frac{B}{p} \text{ where } p \text{ is the parallax.}$$

So 61 Cygni is at distance

$$d = \frac{300 \times 10^6}{\frac{0.6}{3600} \times \frac{2\pi}{360}}$$

$$= 103.13 \times 10^{12} \text{km}$$

At about the same time as Bessel was working on his northern-constellation star, a southern-sky observer, Thomas Henderson, who worked at the Cape Observatory in South Africa, was measuring the position of one of the brightest stars in the southern sky, Alpha Centauri. He found a displacement of 1.6″, showing that this star was much closer than 61 Cygni, even though its proper motion was smaller, 3.8″ per year. Since the sum is the same as above with 1.6 in place of 0.6, all we need to do is multiply our previous result by $\frac{0.6}{1.6}$ so we find Alpha Centauri is at less than half the distance, about 38.7×10^{12}km.

As luck would have it, Henderson's choice was almost the nearest star to the Sun. Alpha Centauri has a faint companion, which is even closer to our Sun. This is Proxima, a dim, red star quite invisible to the naked eye, and needing a moderately large telescope, by amateur standards, to see from the southern hemisphere.

Astronomers measure parallax as the angle subtended by the radius of the Earth's orbit at the star, not the diameter as we have used above. Thus the angles measured are half those given, and so is the base line. Thus the parallax of 61 Cygni is 0.3″, Alpha Centauri 0.76″, and of Proxima Centauri, 0.79″.

Astronomers now had an idea of how small the value of parallax was, and it quickly became evident that, by comparison with most stars, those of 61 Cygni and other neighbours were enormous! Because of the enormous distances involved, a new unit of length was needed to produce more manageable figures. The distance unit chosen was the *light year*, the distance light travels in a year. The speed of light in a vacuum is 299 792.5km/s, so in a year light travels $9.460\ 7 \times 10^{12}$km; this is the value of the light year, and is about 63 240 AU.

Since the calculation of stellar distances is derived (for the nearer stars) from parallax, another unit became used for such distances, the *parsec*. This is the distance at which a star would have a parallax of one second of arc, and is equal to 3.261 6 light years. Using this unit, the distance follows very simply from the parallax: you merely take the reciprocal of the parallax in seconds of arc and you have the distance in parsecs. If you like, you can multiply this value by 3.3 to give light years. Very long distances are measured in thousands of parsecs, kiloparsecs (kpc); or millions of parsecs, megaparsecs (Mpc).

When it became clear that parallax was only of use for measuring the distances of the relatively few stars close to the Sun, a new technique was required. Logically, the brightest stars could be assumed to be close to the Sun, so parallax measurements were made of these, with an incredible result. While Alpha Centauri and Sirius (Alpha Canis Majoris), which were the third-brightest and brightest stars in the sky respectively, also proved to be close (Alpha Centauri closest of all, and Sirius about fifth closest at 8.7 parsecs) the second-brightest star, Canopus, had an almost immeasurably small parallax. Using alternative methods to find its distance, which is about 200 parsecs, Canopus has a parallax of only 0.005″. The implication is amazing: since light intensity diminishes as the square of the distance from the source, Canopus may be some 75 000 times more luminous than the Sun!

So the apparent brightness of stars is most unreliable as a distance indicator. In fact in a list of the twenty brightest stars and one of the twenty nearest, only Alpha Centauri, Sirius and Procyon are found in both. There are nearly forty stars closer than Altair, the next close, bright star.

Parallax measurements are of limited use once the distances involved reach 100 parsecs or so. The distances beyond this have to be deduced from even more flimsy evidence, which is why many of the figures involved are frequently revised (usually upwards—the Universe is always bigger than we think, it seems).

Since the whole Galaxy is in motion, and since the stars are likely to move relative to some arbitrary point at about the same velocity, we can use the effects of the motion of the Sun itself, the change in perspective of near and distant stars. Herschel discovered that a group of stars in the general area of the bright star Vega, Alpha Lyrae, apparently were moving away from a common point close to this star. The effect, he deduced, was due to the Sun's motion in space towards that point, perspective causing the stars to be spread further apart as we approached them.

Once the velocity of the Sun is known by measuring the displacement of stars of measurable parallax, the base line for a parallax measurement can be extended considerably; we merely observe the change in the star's position over many years, using the distance covered by the Sun during that time as the base line. Of course, we have to correct for the motion of the star itself, and here another tool comes into its own, the spectroscope. The speed at which a star is moving towards or away from the Sun can be worked out from the amount by which the characteristic spectral lines in the light from the star are displaced relative to a source at rest.

This displacement of the spectral lines is caused by a shortening of the wavelengths of light from a body approaching the Sun at high speed, displacing spectral lines to the blue end of the spectrum, or the famous red shift of spectral lines of receding bodies. A similar shortening and lengthening of wavelength occurs with sound waves emitted by a moving body, causing, for example, the note of a train whistle to drop in pitch as it rushes past. This is known as the *Doppler effect*.

Once we know the speed at which the star is receding or approaching, and have measured its proper motion, which is the component of that star's velocity at right angles to the line of sight, we can work out its probable distance by assuming that its real velocity will not be much different from those of all other stars.

Another approximation is useful with binary stars. If the most Sun-like of the components of a binary is assumed to have a mass similar to the Sun's, then from Kepler's third Law (see page 21) the separation of the two stars can be worked out. The angular separation and the inclination of the system can be observed and measured, and hence the distance of the system inferred.

Fortunately, the spectra of stars have characteristics which allow their actual luminosity to be deduced within a reasonable margin of error. By comparing the calculated brightness with the observed brightness the distance can be calculated from the inverse square law of luminous intensity. This method is very important, and it is also very important to know the actual brightness of stars because of the information that can be deduced concerning their mass and size from this.

As well as the magnitude, which is the apparent brightness of a star (both visual and photographic magnitudes are used, referring to the brightness as detected by the eye on one hand, and the image on a photographic plate on the other), astronomers use *absolute magnitude*, which is the brightness of the star as it would appear if brought to a standard distance of 10 parsecs from Earth.

In both cases magnitudes are measured on an arbitrary scale

in which a difference of five magnitudes represents a hundred times the brightness. The smaller the magnitude number, the brighter the star. Thus Vega is visual magnitude 0.03, about 100 times brighter than a magnitude 6 star. Sirius is –1.44, so bright that it has a magnitude less than zero to maintain this relationship of relative brightness.

It has already been noted that Canopus is very bright at magnitude –0.72, compared with Sirius' –1.44, and yet Canopus is some 200 parsecs distant compared with Sirius' 2.7 parsecs. At 10 parsecs the magnitude of Sirius (i.e. its absolute magnitude) would be +1.41 compared with Canopus which may be as much as –7.5. The Sun at 10 parsecs would be magnitude 4.79, 22 times dimmer than Sirius and 75 000 times less bright than Canopus. But that is nothing! Stars with absolute magnitudes of 15 or dimmer are plentiful, and they are 1.5×10^9 times dimmer than Canopus!

The mass of stars

Newton showed that the force of gravitational attraction between two bodies was proportional to the product of their masses, and inversely proportional to the square of the distance between them, the constant of proportionality, G, being universal. Its value is $6.670 \times 10^{-11} m^3/kg/s^2$ which has been determined by laboratory experiments that are simple in principle.

When one of the two masses is a planet and the other an object on that planet, the distance of the object will not vary much from the centre of gravity of the planet and hence all the factors in the relationship are virtually constant. The force any mass exerts, called weight, is therefore a constant value for each unit of mass on any one planet. On Earth, this force causes an acceleration of $9.81 \, m/s^2$ on a freely falling body, no matter what its mass. This constant, usually called g, is also simple to determine.

From this we can say that a mass of 1kg at a distance of about 6378km from the centre of the Earth is subject to an accelerating force of 9.81 newtons (1 newton is a force of about 100 gram weight). But force $F = \frac{G\ M_1\ M_2}{d^2}$ where M_1 and M_2 are the masses of the Earth and the object, and d is the distance between them, the radius of the Earth in this case. So the mass of the Earth can be determined by rearranging to give

$$M_1 = \frac{F\ d^2}{G\ M_2}$$

$$\text{so } M_1 = \frac{9.81 \times (6378 \times 10^3)^2}{6.67 \times 10^{-11} \times 1} \text{ kg}$$

which gives the mass of the Earth as 5.98×10^{24}kg.

From this same law of Newton's, we can see that the force between two bodies in orbit around a common centre of gravity produces an acceleration towards that centre of gravity.

Force is also given by mass times acceleration, according to Newton's laws. So for a planet in orbit around a star, or a satellite around a planet, for which we can assume that the smaller body orbits the centre of the larger (only by measuring with respect to some third point can we tell any different, and in any case the difference in most cases between the true centre of the large body and the common centre of gravity is negligible compared with the distance between the two bodies), the accelerating force is the mass of the orbiting object, M_o, times its angular acceleration. Now since angular acceleration is proportional to the radius of the orbit, d, divided by the square of the orbital period, t, we have:

$F \propto M_o \frac{d}{t^2}$, and, as we saw above,

$F = \frac{G\ M_c\ M_o}{d^2}$, where M_c is the mass of the central body,

so that $\frac{M_c\ M_o}{d^2} = \text{constant} \times \frac{M_o\ d}{t^2}$

or $M_c = \text{constant} \times \frac{d^3}{t^2}$.

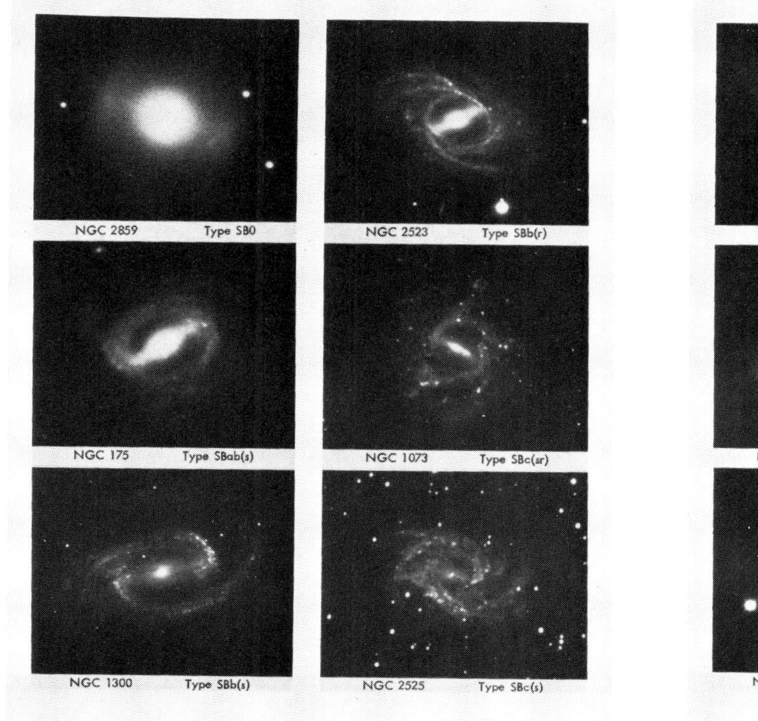

Plate 14: Examples of barred spiral galaxies. Note that the SBO type is almost an elliptical, and the SBc(s) is fairly irregular. *(Photographs: Hale Observatories)*

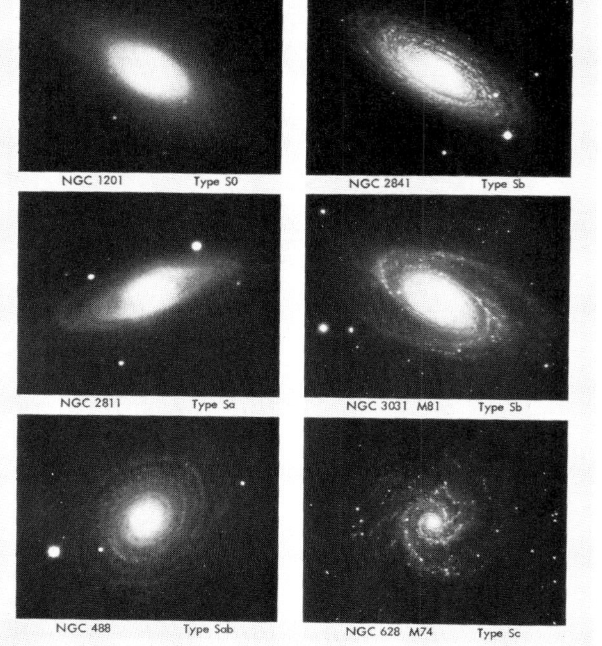

Plate 15: Examples of normal spiral galaxies. *(Photographs: Hale Observatories)*

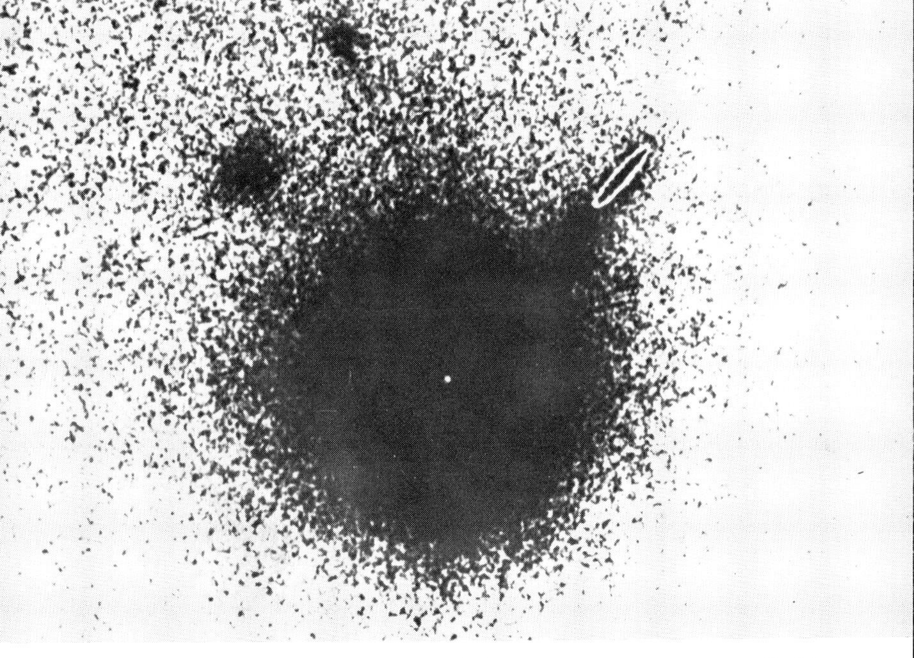

Plate 16: Quasar 3C 273. The white dot in the centre represents component B of the radio source, and the white oval highlights the jet which is component A of the source. *(Photograph: Hale Observatories)*

Plate 17: A cluster of galaxies in Coma Berenices at a distance of some eleven megaparsecs. *(Photograph: Hale Observatories)*

In any solar or planetary system the mass of the central body will be the same for all calculations, so that in the case of the Sun $\frac{d^3}{t^2}$ will be constant for all the planets, which is the third law that Kepler deduced empirically. The important factor is the constant G, which is fixed for anywhere in the universe, at least wherever Newtonian laws of gravity are true.

This means that we can find the ratios of masses of any two bodies, once we know their separation and orbital period. Since $M_o = \text{constant} \times \frac{d^3}{t^2}$, we can evaluate this expression by taking any two systems, say the distance and mean period of the Moon/Earth and the Earth/Sun systems.

In this case suffixes s, m and e refer to the Sun, Moon and Earth respectively:

$$M_s = \text{constant} \times \frac{d_e^3}{t_e^2}$$

$$M_e = \text{constant} \times \frac{d_m^3}{t_m^2}$$

so if we divide one equation by the other we have

$$\frac{M_s}{M_e} = \left(\frac{d_e}{d_m}\right)^3 \times \left(\frac{t_m}{t_e}\right)^2$$

So the mass of the Sun can be determined by the combination of a simple observation and a laboratory experiment to be:

$$\frac{M_s}{M_e} = \left(\frac{149.6 \times 10^6}{0.384 \times 10^6}\right)^3 \times \left(\frac{27.32}{365.25}\right)^2 = 330\,000$$

This is the number of Earth masses in the Sun, so the Sun therefore has a mass of 1.99×10^{30}kg.

The relationship of two bodies of comparable mass rotating about a common centre of gravity is simply that $M_1 + M_2 = \text{constant} \times \frac{d^3}{t^2}$. Kepler missed this more accurate statement of his third law simply because the masses of the planets are negligible compared with the Sun. But we have dealt with these laws at such length to show that they are not particularly difficult, and yet they release a floodgate of astronomical knowledge.

For any pair of mutually rotating bodies in the Universe, with masses M_1^* and M_2^*, we can use the above expression and compare it with the Earth/Sun system. So by observing a binary star to determine its orbital period, t^*, working out its distance by parallax or spectroscopic methods and translating this into actual separation, d^*, from observation of the angular separation of the two components, we know that:

$$\frac{M_s + M_e}{M_1^* + M_2^*} = \left(\frac{d_{s-e}}{d^*}\right)^3 \times \left(\frac{t^*}{t_{s-e}}\right)^2$$

Since the Earth adds less than one-thousandth of one per cent to the total mass of the Earth/Sun system, we can ignore M_e. We can also express all masses in terms of solar masses (mass of Sun = 1) and distances in terms of astronomical units. Then the expression simplifies to:

$$M_1^* + M_2^* = \frac{d^{*3}}{t^{*2}}$$

—a veritable key to the cosmos!

The total masses of binary-star systems can now be calculated using only the observed separations, parallax and period since the parallax in radians is the inverse of the distance to the star in AU, and the conversion factor also applies to the separation, hence

$$M_1^* + M_2^* = \frac{d^3}{p^3 \times t^2}$$

where separation d and parallax p are in seconds of arc.

Thus the nearest bright star, Alpha Centauri, which is a binary having a maximum separation of 18″, parallax 0.75″ and period 80 years, has a combined mass of 2.16 solar masses. Both stars are similar to each other and to the Sun in class (see below) and brightness, so it can reasonably be assumed that they are each about the same mass as the Sun.

The range of star masses is quite small, compared with the vast differences in brightness, size or density, and in fact below about a few per cent of the mass of the Sun it is doubtful whether a large companion can be called a star, and should then probably be classified as a large planet.

The classes of stars

The next weapon in the astronomer's armoury is the spectroscope. This splits the light from the star into a spectrum of different wavelengths, and allows the study of bright lines caused by the emission of energy from the atoms of particular elements at characteristic wavelengths. This allows an enormous amount of data to be deduced about the nature of the star, including its chemical composition and temperature. When the stars were first examined by spectroscope it was clear that certain types of star produced certain characteristic lines in their spectra, due to the emission of energy at different wavelengths characteristic of excitation at different temperatures. Just as a very hot piece of iron emits a characteristic red colour, and the sodium in a sodium lamp emits a familiar orange, so other elements give out wavelengths that can be identified by comparison with the spectrum of a known substance excited until it emits visible energy. The stars were divided into classes of temperature ranges, which were A, B, C, D etc. starting at the highest temperatures.

Later, as astronomers discovered more about the stars, the classes were changed and subdivided, generally keeping only some of the same letters as before and revising the order to suit the known temperatures, until the order became W, O, B, A, F, G, K, M, R, N, S. (This can be remembered with the famous mnemonic: 'Wow! Oh be a fine girl, kiss me right now Sweetie!')

The majority of stars fall in the classes between B, the hottest blue/white stars, and M, the 'cool' red stars. The Sun is a class

G star (class G2V, to be precise), which is average—luckily for us!

At the extremes of the scale of stellar classes are the fantastically high-temperature class W stars which, together with class O, were discovered by two French astronomers, Wolf and Rayet, after whom the two classes of star are named. Wolf-Rayet stars of class W have wide emission bands in their spectra, and are believed to have temperatures of over 30 000 to 40 000°C. Stars can hardly be expected to last long in this condition, and type W stars must quickly (on the cosmological time-scale) change to a cooler type of star. There are very few class W (or O) stars compared with classes B to M. The most easily found examples of class O stars are the two outer stars of Orion's Belt, the distinctive line of three stars across the middle of the constellation. They are both double stars (see below) and easily seen in the months of December, January or February. The constellation of Orion as a whole is shown in plate 2.

At the other extreme are the dim red class R, N and S stars, which are usually variable in brightness and have surface temperatures of only 2500°C. In the sky near the Great Nebula in Andromeda, which is the nearest major galaxy to our own, is a pair of magnitude-4 stars just west of the line drawn towards the pole from the Andromeda side of the Great Square in Pegasus. These two stars form a little triangle with a third, R Andromedae, only 4° east of the Great Nebula itself, and a little south. This variable star R is class Se long-period variable, ranging in brightness from magnitude 5, which is just visible to the naked eye, or easily seen in binoculars, to less than magnitude 15, which is visible only in a telescope of over 250mm aperture. We shall consider variable stars later, but it is interesting to note that many of the irregular and very-long-period variables (R Andromedae takes about 13 months to go through a cycle of brightness) are of classes at the red end of the spectrum such as M to S.

The principle characteristics of the main classes of stars are summarised in the table.

Class	*Colour*	*Surface Temperature (°C)*	*Examples*
W	White	40 000	
O	White	30 000	Zeta and Delta Orionis (Orion's Belt)
B	Blue/white	12 000	Rigel, Achernar
A	White	8 000	Sirius, Deneb, Altair
F	Yellow/ white	6 000	Canopus, Procyon, Polaris
G	Yellow	4 000	The Sun, Capella, Alpha Centauri A
K	Orange	3 500	Arcturus, Aldebaran Alpha Centauri B, Epsilon Eridani
M	Orange/ red	2 500	Betelgeuse, Antares, Mira
R, N, S	Red	2 300	

The spectrum of a star gives information not only about temperature and chemical composition. As more data became available on star masses and absolute magnitudes, using the methods discussed above, it became evident that there was a relationship between some of the stars' characteristics. If the main spectral types of star are plotted in order of temperature (OBAFGKM) against absolute magnitude, a distinctive pattern is observed. This is the Hertzsprung–Russel diagram, named after a Danish and an American astronomer who first drew attention to this relationship. Although there were many stars which fell outside the 'main sequence' of stars on the diagram, it enabled the absolute magnitude of many very distant stars to be determined.

Another piece of vital information fits into the puzzle of determining the nature of the distant stars. One well-known type of variable star, the Cepheid variables (named after the bright component of the double star Delta Cephei which varies between about 3.6 to 4.3 magnitude over a cycle of 5.37 days) proved to have a period precisely related to their absolute brightness. This is believed to be due to expansion and contraction of the star and, the more massive the star, the longer is its period of variation and the higher its luminosity. This type

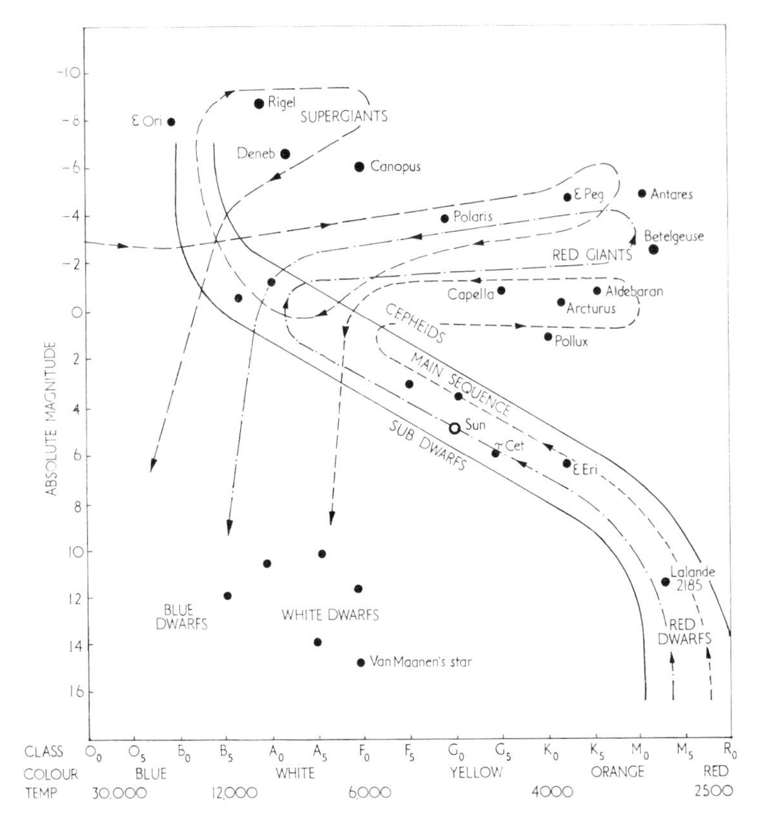

19 The Hertzsprung–Russel diagram of stellar class and absolute magnitude, showing the 'main sequence', the position of most stars on the diagram, and other selected stars. The broken lines show possible evolutionary paths through the diagram for different types of protostars.

of star is most useful in the determination of the distances of the nearby galaxies and clusters of stars, since, if a Cepheid variable can be found in the group and its period and apparent brightness measured, the absolute magnitude can be determined and hence the distance.

The next important relationship between the different characteristics was discovered in 1924 by Sir Arthur Eddington, a British astrophysicist who was one of the pioneers of the

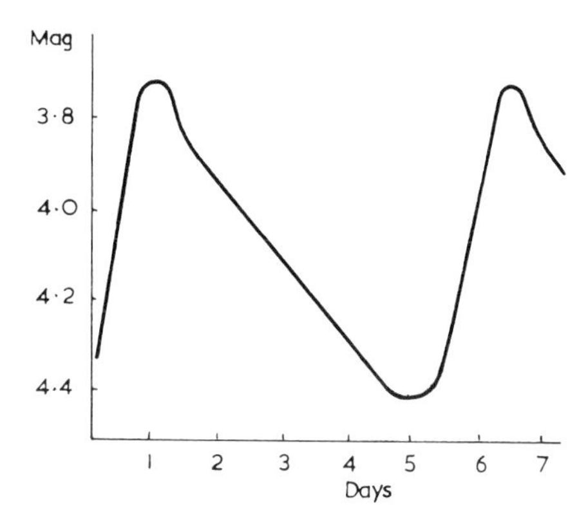

20 Light curve of Delta Cephei, the type star of the Cepheid variables.

modern theories of stellar evolution. Eddington was studying the luminosity of binary stars by spectroscopic methods, and, since the mass could be determined as we saw earlier, he was able to determine both mass and luminosity accurately. Strictly, the term *absolute bolometric magnitude* should be used, which means the absolute visual magnitude corrected to include infrared and ultraviolet radiations that have no effect on the visual brightness.

For the stars of the more typical absolute magnitudes and masses (magnitudes –1 to 10, and 0.25 to 4 solar masses) the relationship between mass and absolute bolometric magnitude is quite consistent.

The size of the stars

So now we can find out how far, how bright and how massive the stars are. What about size? The range of stellar masses is nothing like as great as the range of luminosities. Can we infer the range of sizes from this? The radiation from a surface increases as the fourth power of the temperature, T, and we have now found a number of ways to determine temperature. But if we take the class as the guide it will serve as an example. If Q is the total energy per unit area, proportional to T^4, and L is the luminosity (Sun = 1), we can say that the total energy emitted is proportional to luminosity, and in turn that this is proportional to the fourth power of temperature and to the area of the star. The area of a sphere is proportional to the square of the diameter, D, so if we use solar quantities as a comparison we can say:

$$\frac{D^2\ T^4}{D_s^2\ T_s^4} = \frac{L}{L_s}$$

where the suffix s denotes the Sun's characteristics.

The diameter of the star is therefore

$$D = \frac{D_s\ T_s^2}{T^2} \sqrt{\frac{L}{L_s}},$$

and if we set $D_s = 1$ the answer is in solar diameters. To find the comparison in energy emission, we have to take the difference of absolute brightness. Since a difference of 1 in magnitude represents a difference in brightness of $\sqrt[5]{100}$ = 2.511 886 4 we must raise this figure by a power equal to the difference in magnitudes thus:

$$L = (\sqrt[5]{100})^{\Delta M}$$

where ΔM is the absolute magnitude difference.

This is near enough to $2.5^{\Delta M}$. So if we take Arcturus, assuming its surface temperature to be 4100°C compared with the Sun's 5500°C, and its absolute magnitude to be –0.3 compared with the Sun's 4.8, we can say; $\frac{L_a}{L_s} = 2.5^{5.1} = 107$.

So Arcturus is over 100 times more luminous than the Sun. Its diameter is then given by:

$$\begin{aligned} D_a &= \frac{T_s^2}{T_a^2} \times \sqrt{107} \\ &= \frac{5500^2}{4100^2} \times 10.344 \\ &= 18.6 \end{aligned}$$

Arcturus is about 19 times the Sun's diameter.

Even the closest stars are too far away to show anything but a point of light even at the highest magnifications, although optical effects in any telescope will give a spurious increase in size to the images of stars, proportional to their brightness. From a hypothetical planet of Proxima Centauri, for example, at a distance of $4.2 \times 9.460\,7 \times 10^{12}$km $= 39.734\,9 \times 10^{12}$km from the Sun, the Sun's diameter of 1 392 000km would subtend an angle of two-millionths of a degree, or about seven-thousandths of a second of arc. The magnification of any Earth-based telescope is limited by the size of the optical surfaces which can be achieved. To resolve this tiny angle, the diameter of the telescope mirror would have to be at least 16.5 metres, over three times as big as the world's largest existing or planned telescopes—and even then it would be at the theoretical limit of resolution. But such a telescope would have to be mounted in outer space, since the Earth's atmosphere will distort the image in any telescope once the magnification is increased enough. The limit would be well below the magnification required to detect such a small disc. Tiny Pluto, say 6000km in diameter, at a closest distance of 4273 million kilometres, subtends an angle of about 0.3 arc sec, also quite immeasurable at present. To see our Sun from the nearest star is comparable to trying to see a 20mm diameter coin at a distance of 570km!

Interferometry

However, there are indirect optical ways of measuring the size of a star. A technique was developed by the American physicist Michelson to detect very close binary stars, based on interferometry. This makes use of the fact that light waves from two sources may be displaced relative to each other to cause alternate reinforcement and diminution of the aggregate light wave. If the light from two points on the telescope objective is brought to a common focus, a distinctive pattern of 'fringes' is produced. If the source of this light is in fact two fairly widely separated points, the light will take a finite time to cover the extra distance between the two points before setting off on its years of journeying across space to Earth, and all the time this 'phase difference' will be retained. The result is a change in the interference pattern when seen in the interferometer.

The astronomer Pease found that not only could close binary stars be detected in this way, but that some stars were so large that the interference fringes due to light from different parts of their surface could be detected. The interferometer confirmed the incredible results of calculations by the luminosity method that some super-giant stars existed that were 1000 or 2000 times the diameter of our Sun. In fact our own Sun is now classed as a G2V 'dwarf'.

Near the bright star Capella in Auriga (found to the north of Orion by extending a line from Rigel through Gamma Orionis,

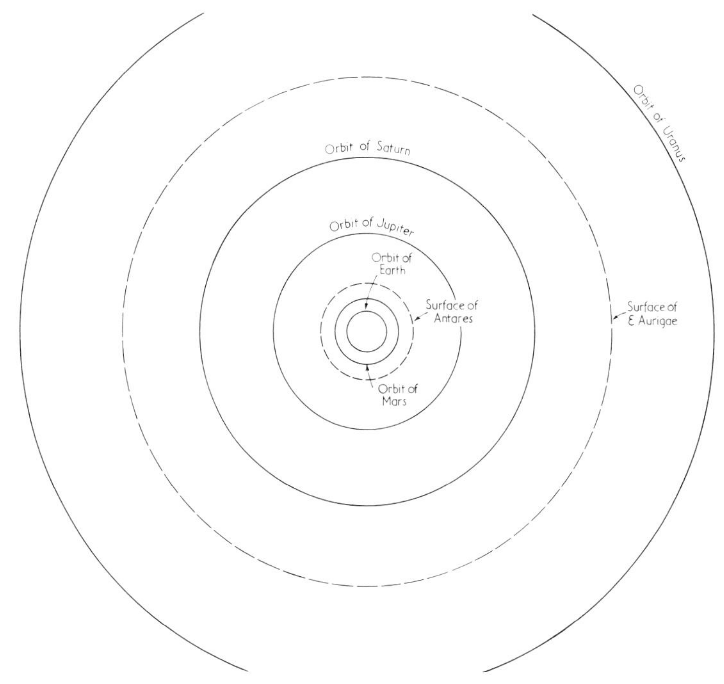

21 Supergiant stars compared with the orbits of the planets. The giants are composed of very tenuous gases.

the star forming the western corner of the main constellation), we find an elongated triangle of third-magnitude stars, called 'the Kids'. The vertex of the triangle, near Capella, is Epsilon Aurigae. This is a most interesting star. It varies regularly over a long period, 27 years, suggesting that it has a dark companion star which obscures some of the bright star's light from Earth. Many such eclipsing binaries are known, but few are so widely separated to give such a long period. There is much speculation about this dark companion, including the idea that it might be a black hole (which we will examine later). But the bright star Epsilon Aurigae A itself is possibly over 2000 times the diameter of the Sun, greater than the diameter of Saturn's orbit!

Binary stars can also be detected by the red-shift in the spectrum of one body with respect to the other, indicating that one is receding, or a blue shift, indicating the star is approaching. If the shift regularly varies from red to blue it clearly indicates an orbiting body. From this the radial velocities of the orbiting star can be worked out and, if the stars are in line of sight so that the two bodies pass behind one another, the variation in the light curve from the two can be used to translate the radial velocity into a measurement of the sizes of the two bodies. Once again, the results can be compared with interferometric or luminosity calculations. Such comparisons have given astronomers confirmation of the limits of accuracy of the different methods.

We have seen how small the diameter of a star appears from the Earth, even the supergiant Epsilon Aurigae would only be 19 arc sec diameter if it were at the same distance as Proxima Centauri. But Epsilon Aurigae is 1000 parsecs distant, so it is only 0.024 arc sec in apparent diameter.

The twinkle method

The Earth's atmosphere is continually in violent motion due to different temperature layers and streams in the atmosphere. The refractive index of any transparent medium varies to some extent with temperature, so that the effect of the Earth's atmosphere is to cause rapid and continuous slight variations in its refractive index. This causes the tiny points of light to 'dance' when seen in a telescope, and on many nights the stars' position varies even as seen by the naked eye, causing the twinkling which we all admire on a cold, clear night, but which is a menace to observers.

Planets, on the other hand, look like very bright stars to the naked eye, but rarely twinkle, since it requires a very turbulent atmosphere indeed to cause the planetary disc to move in apparent position by more than its own diameter. This can occur when the planet is very low in the sky, and it may occasionally occur with the smaller apparent discs of Uranus and (rarely) Mars. But, clearly, twinkling is a function of the apparent size of the body in our sky.

This has been used recently for a new method of estimating star diameters. If the star image is photographed rapidly several times in succession on the same film, twinkling will cause several images of the star to be recorded. For a given set of sky conditions, the patterns obtained can be compared for different stars to indicate the relative sizes of their apparent discs. Workers at the Paris observatory recently designed a computer method of examining 'speckle interferometry' photographs taken by the Kitt Peak and Palomar observatories, and found sizes of 0.018 and 0.054 arc sec for two stars. This was the first time stellar diameters had been measured directly by an optical measurement. This method has also been used to check measurements made by older techniques, enabling these to be refined.

We can now use these methods to investigate the peculiar behaviour of the bright red star Betelgeuse in the northwest corner of the main part of Orion. Betelgeuse varies in brightness irregularly from 0.06 to 0.75 magnitude, averaging 0.41, and is a class M2 star, absolute magnitude –5.6. From the mass/luminosity relationship we find that it has a mass some 18 times that of the Sun.

From the difference in absolute magnitude compared with

the Sun, and the lower temperature, about 3000°C, we find that the size is about 390 solar diameters. A red giant indeed. But astronomers have found spectral shifts which suggest that the surface of the star recedes and approaches with the light variation, which suggests that Betelgeuse is actually changing in diameter. From the ratio of the observed magnitudes, this corresponds to diameter variations of from about 290 to 450 solar diameters.

A star with 390 times the diameter of the Sun has about 59 million times the volume of the Sun. Since the mass is only 18 times that of the Sun, the density of Betelgeuse is only 30 millionths that of the Sun, giving the specific gravity as 43×10^{-6}, which for most purposes is a very good vacuum! Red supergiants such as Betelgeuse do not fit into the main sequence of stars in the Hertzsprung–Russel diagram. The reactions deep inside these stars must be very different from those in the Sun. At these low temperatures, protons can collide with hydrogen nuclei to form helium directly.

Clearly, Betelgeuse must be very tenuous simply to explain its irregular changes in size. These may be due to internal pressures building up, causing the star to expand until the nature of the internal reactions are changed, resulting in a resumption of the 'normal' behaviour.

Another supergiant in Auriga, Zeta Aurigae, is also in the triangle of the 'Kids'. Zeta is the western star of the two in the base of the triangle, and the faintest of the three. This is a K4 star which has a spectroscopic companion of class B8 (i.e. a very luminous star) which passes into eclipse behind the supergiant every 2½ years. The light curve of this event shows the gradual fading of light from the small star and its gradual reappearance, both lasting several weeks, as the B8 star shines through the outer regions of the tenuous K4 giant.

At the other end of the scale are the dwarf stars. Although the Sun qualifies as a G-type dwarf, its specific gravity is only 1.409—not much more dense than water.

The famous companion of Sirius caused a stir in the astronomical world when it was first seen by the American telescope-maker, Clark, in 1862. The companion had been predicted by Friedrich Bessel almost 30 years earlier from studies of the variations in the proper motion of Sirius itself. The star was being perturbed on either side of its path through the heavens by a tiny amount almost exactly every 50 years. The mass of the companion was clearly of the same order as that of the primary, but being quite invisible it was certainly very dim. When Clark saw Sirius B for the first time during a test on a new telescope it was found, to everyone's surprise, to be fairly faint but brilliant white, indicating a small, very hot body, with a surface temperature near 10 000°C, not a massive, dim red star as expected. With an absolute magnitude of 11.54, the size of this A5-class star can be found as above, but this time the mass is known from Kepler's Laws to be about the same as our Sun. The size of Sirius B is now believed to be about 24 000 to 30 000km diameter, smaller than Uranus! If the mass is the same as the Sun's, the density of this star is some 26 000 times that of the Sun. A sample measuring one cubic centimetre from the Sun would weigh 1.4 grams on Earth. The same volume from Sirius B would weigh some 36 kilograms.

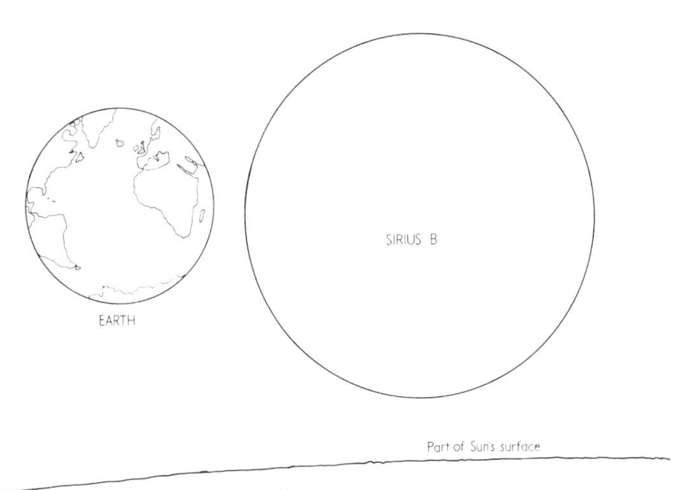

22 The white dwarf Sirius B compared with the Earth and part of the Sun.

White dwarf stars have been found with sizes down to about the same as Mars, and it is now believed that stars may get very much denser even than this. But we shall return to this subject later when considering the life of stars.

More variable stars

Cepheid variables (page 109) were important in determining long distances in space, and eclipsing binary stars proved a valuable clue to finding the sizes of stars, as we saw above. The visible effect of the eclipsing binary is to produce a regular dimming in the light output from the star as the dimmer companion passes in front of the bright star. The companion must be of comparable size with that of the bright star. Our own Sun, when observed from near a star such as Regulus which is almost in the plane of the Solar System, would dim by a very tiny amount as Jupiter made a transit of the Sun's disc; but it would amount to a variation of a small fraction of one per cent (and the Sun would be a miserable magnitude 7 star, invisible to the naked eye!). But comparably delicate magnitude measurements may soon provide evidence of planets circling other stars.

Perhaps the best known eclipsing binary star is Algol, Beta Persei, which was well known to early Arab astronomers, who called it the Demon Star, and which was held to be the eye of Medusa the Gorgon in the constellation of Perseus by the Greeks. Perseus is found just about overhead from moderate northern latitudes at close to midnight at the end of November. It is best recognised as a gently curving arc of four stars between the W of Cassiopeia and the first-magnitude star Capella in Auriga. South of the brightest star, Alpha, in this arc of stars are two stars, the nearest (and usually the brightest) being Algol. The second star, Rho, is an M2-class star which is also variable, but irregularly, similar to Betelgeuse in Orion. However Rho is usually magnitude 3.3. Algol varies with clockwork regularity between magnitude 2.3, only a little less bright than Alpha, for most of the time, to 3.5, about the same

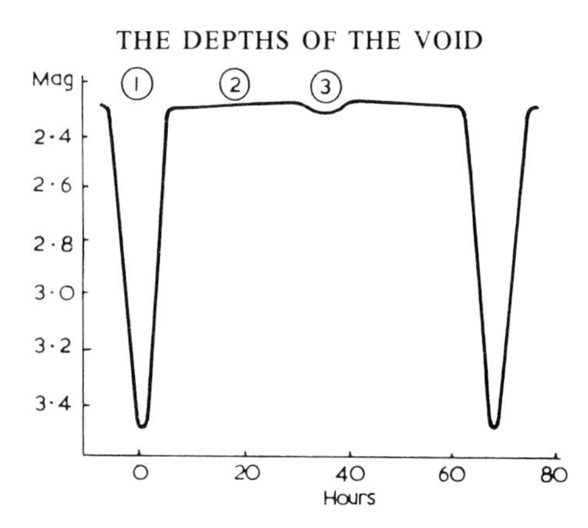

23A Light curve of Beta Persei, Algol, the type star of the Algoloid eclipsing variables (see chart 2).

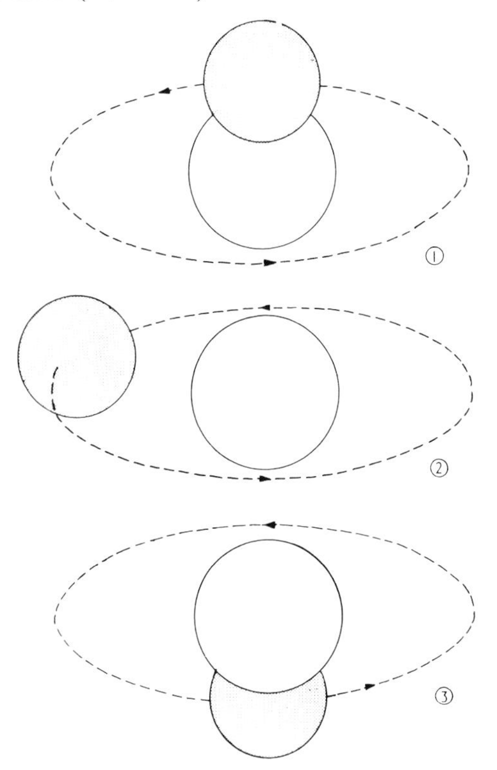

23B Principle of the eclipsing binary. It is now known that the bright component of Algol is a spectroscopic binary.

as Rho is normally. This change in magnitude occurs every 2.87 days, and the brightness falls from maximum to minimum in 5 hours and returns to normal brightness in another 5 hours. A tiny drop in brightness, of 0.05 magnitude, also occurs half way through the period between the major drops in brightness. This star is visible for much of the year so is well worth watching. The explanation for Algol's variation is simply that we see the light from two components of a binary system, but one is very much dimmer than the other. Their orbits are such that the dimmer body partially eclipses the brighter, that is, it passes in front of the brighter one as seen from Earth, causing a dramatic drop in the light from the pair; then, 1.435 days later, the brighter one obscures part of the disc of the dimmer star, causing a tiny loss of the total light from the pair.

Many eclipsing binaries have two bright components in orbit about each other. They can also eclipse each other, but the loss of light is not so pronounced as with stars like Algol, where the two main components (Algol appears to have a third component) have a ratio of brightness of about 15:1. Some of the bright eclipsing variables show a pronounced dip in brightness twice in one cycle. This is believed to be caused by an apparent change of shape of the stars as seen from the direction of Earth. The two components are so close together in their orbits about each other that their mutual gravitational attraction pulls the stars into egg shapes. The distorted stars are locked in synchronous rotation, just like the Moon is in its

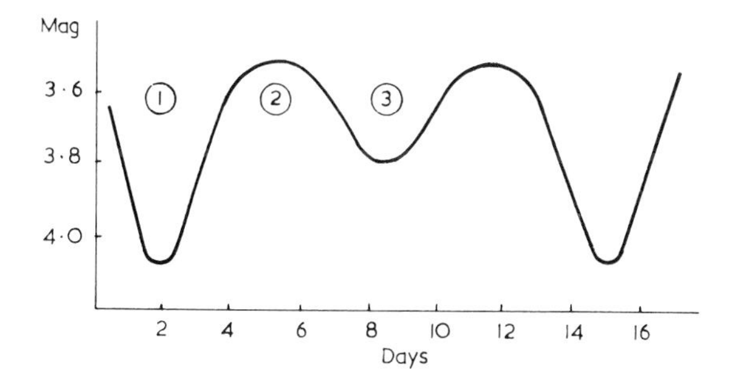

24 Light curve of Beta Lyrae (see chart 6).

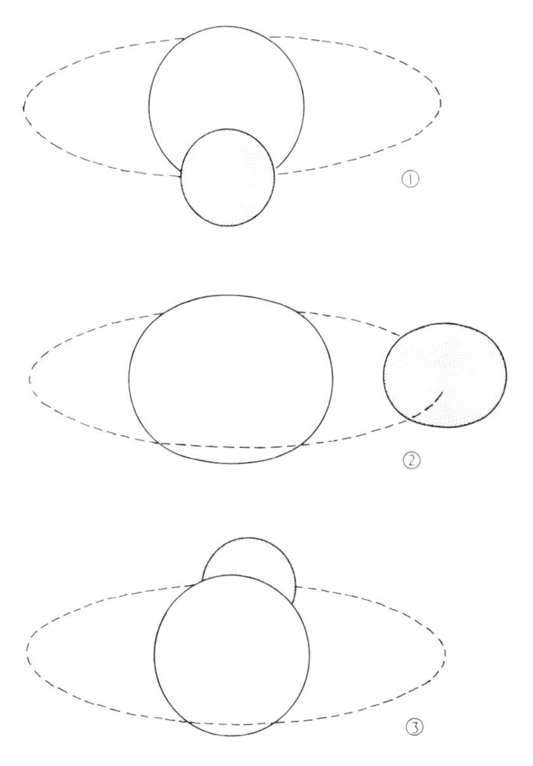

25 Although the components of a Beta Lyrid variable may be of slightly different brightness, the variation is mainly due to the difference in the stars' surface presented to the observer, because of the oval shape of the components and their mutual occultations.

orbit round the Earth, so that we see first the side view of two egg-shaped stars, then, as they partially obscure each other they present an approximately circular profile to our view, causing a greater reduction in the total light output than that due to the eclipse alone.

Binary stars are interesting because stars like Zeta Aurigae give us clues about the atmospheres of giant stars, and they are valuable milestones in the galaxy, but the mechanism of their light variation is simple and does not tell us much about the stars.

Variables such as the irregular class-M red supergiants, on the other hand, are very important because the variations in their light output are due to the nuclear reactions going on deep inside the star, and hence provide valuable clues to the nature of the stars, and, more important, clues to the life cycles of stars.

Variable stars fall into several distinct categories, each usually named after the first or best-known example of this type of star. For example, Spica, Alpha Virginis, is an Algoloid eclipsing binary, although it has only a small range of magnitudes, about 0.9 to 1 in 4 days. Bright eclipsing variables where tidal distortion causes an increased light variation are called Beta Lyrae types after the third-magnitude star south of Vega which varies between 3.4 and 3.8 in the first half cycle, and 3.4 and 4.1 in the second, the complete cycle taking just under 13 days.

In general there are two types of variables where, it appears, the star itself varies in brightness. These are the regular or irregular variables where the changes in brightness continue to occur over long or short periods, and the eruptive stars which occasionally flare into sudden brightness (or suffer sudden and unpredictable drops in light output) or which burst into sudden activity so that a star appears in the sky where none was seen before.

Exploding stars

The latter types, called *novae*, or new stars, are occasionally so bright that the sudden and tremendous change in brightness can only mean an explosion of the entire star of such violence that the star must be completely destroyed as a result. Such stars are called supernovae and are very rare. Most supernovae are found in other galaxies, simply because from Earth we can see only a small proportion of our own Galaxy, whereas we can see millions of other galaxies.

The fact that a bright point of light can appear in the photograph of a distant galaxy and have a light output equal to that of all the other stars in that galaxy, while the galaxy itself in a telescope appears as only a tiny smudge of light, indicates the violence and magnitude of the explosion. The eruption of stars into novae or supernovae clearly has an important position in the overall life cycle of stars, and whenever such an event occurs astronomers study the star in great detail to find out how the outburst fades away, what the spectrum of the novae shows, and what can be found out about the star before the outburst occurred.

Supernovae occur about once per century in a galaxy, but they tend to occur near the central plain of a galaxy. In our own Galaxy most are obscured by the interstellar dust clouds. In fact no supernova has occured 'locally' since the invention of the telescope.

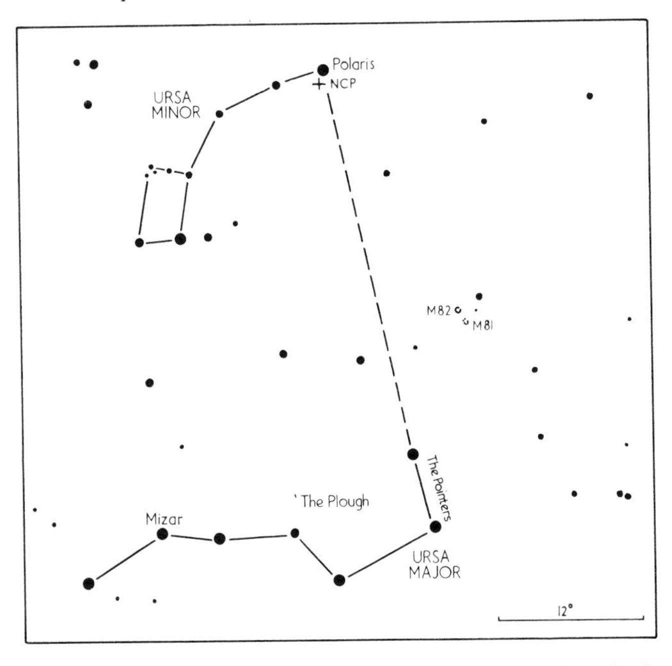

26 Star chart 1. The 'Plough' and the north celestial pole. The bar in the bottom right corner represents a 100mm rule at arm's length (about 12°) in this and the other charts. Double stars include Mizar and Polaris. M81 and M82 are part of a cluster of galaxies.

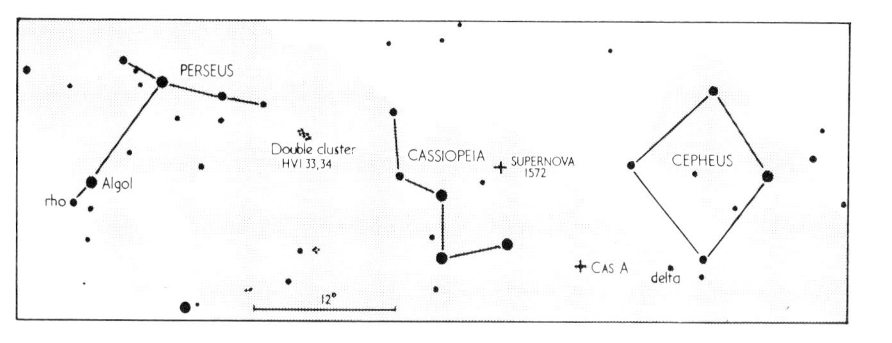

27 Star chart 2. Cassiopeia is found on the opposite side of the north celestial pole to the Plough. Tycho's star (the supernova of 1572) and Cassiopeia A are features of the constellation. Variable Rho Persei is usually a good check for the brightness of eclipsing binary Beta (Algol). Delta Cephei is a double, the brighter component of which is the type star for Cepheid variables. The double galactic cluster HVI 33 and 34 can be seen with the naked eye, yet was missed by Messier.

Novae are less violent in nature, although if our Sun became a nova it would be the end of the planets as we know them. If a nova occurs relatively close to the Sun they can appear quite bright, as did the nova in Cygnus in the summer of 1975 which for a few days quite changed the appearance of the constellation and became almost as bright as Deneb. In the case of a nova, the star is believed to produce an enormous burst of energy and to shed an appreciable proportion of its mass in a very short time, but the star is left in existence. Some stars even behave like a nova several times. These so-called recurrent novae undergo pronounced and rapid changes in brightness characteristic of a nova, usually irregularly. An example is T Corona Borealis which varies between magnitudes 2 and 10.8.

Another important type of variable is the Mira Ceti type. This is a long-period semi-regular variable type that reaches maximum brightness over periods of the order of a year. The actual maximum magnitude each cycle tends to vary. These are usually M-class stars and their behaviour is similar to that of Betelgeuse, in that they are expanding at a tremendous rate.

But the range of magnitudes can be very dramatic. Mira itself, in the constellation of Cetus, is quite difficult to find unless it is near a maximum. But its range of magnitudes, 1 to 10, represents a change in brightness of nearly 4000.

A common factor in many of the variables which have the largest brightness ranges, and the majority of novae, is that they are very red. They appear to represent the dying stage of stars, and whether they explode, or develop the instability of a variable such as Betelgeuse, or simply expire probably depends upon their initial size.

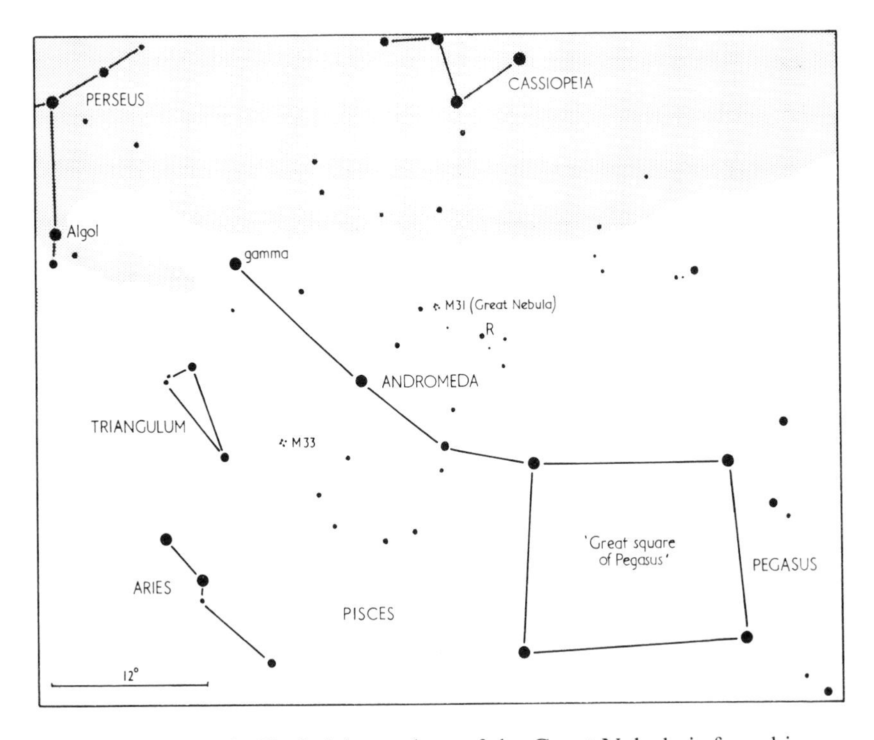

28 Star chart 3. The bright nucleus of the Great Nebula is found in Andromeda as shown. The galaxy itself is long enough to cover about 3°, six times the angular size of the Moon. Another galaxy of our 'Local Cluster', M33, is found in this part of the sky, near Triangulum. Gamma Andromedae is a beautiful blue and amber double star.

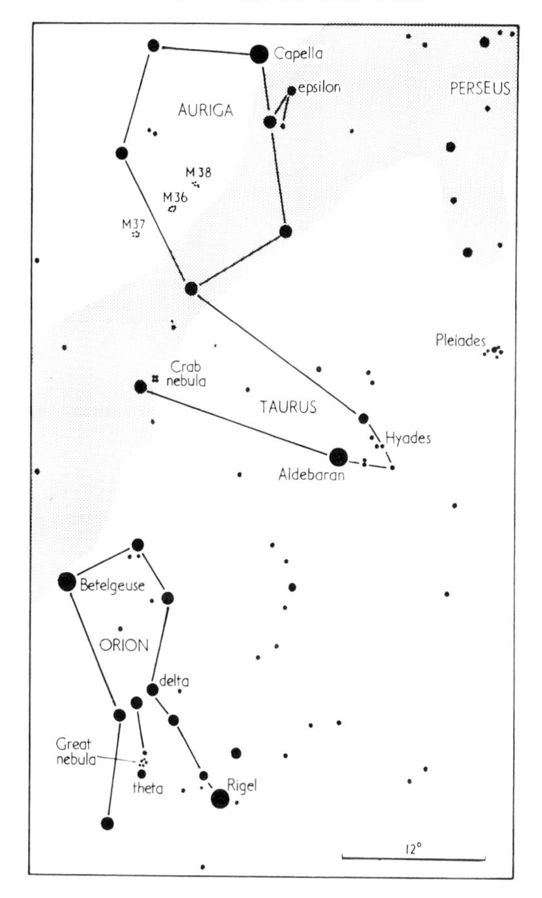

29 Star chart 4. The famous Crab nebula, near Zeta Tauri, and the Great Nebula in Orion are among the most important astronomical objects. Naked-eye clusters like the Pleiades and Hyades are full of newly-formed stars. Three clusters in Auriga, M36, 37 and 38 are worth finding also. One of the largest known supergiants, and possible black-hole companion Epsilon Aurigae, is easily seen close to Capella. Epsilon is an irregular variable. Another is giant Betelgeuse in Orion. Compare this red star with Aldebaran.

The life of stars

We shall see later how astronomers believe the Universe itself is evolving. It is believed that a state was reached when space

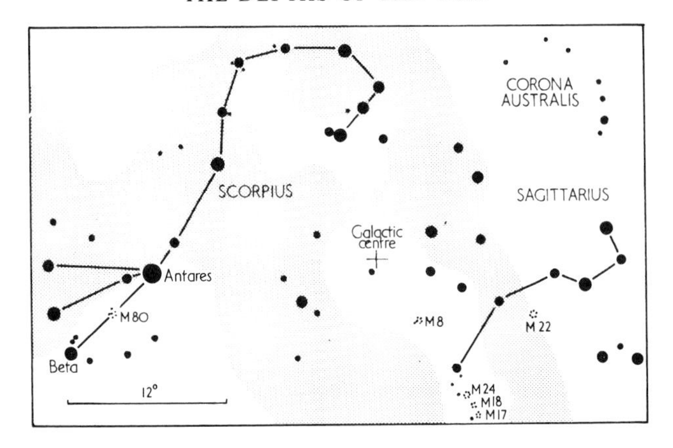

30 Star chart 5. The galactic centre is in Sagittarius. There are many clusters and nebulae in this region. M8 (the Lagoon Nebula) and M17 (the Omega Nebula) are gaseous nebulae. Double stars include Alpha (Antares) and Beta Scorpii.

was full of gases (mostly hydrogen) and the dust of some of the heavier elements, which began to coalesce into the galaxies under the force of gravitational attraction. These huge masses of gas and dust continued to condense into smaller discrete clouds of matter, each something similar to its parent galaxy in shape, a huge globe thousands of millions of kilometres across, even light years in size, rotating about a centre of gravity. This rotation, over millions of years, caused these 'globes' to flatten into disc-shaped clouds. Most of the mass was concentrated at the centre, where gravitational attraction was strongest, until the enormous pressures developed at the centre caused the mass to heat up. Eventually the concentrated masses reached such enormous temperatures and pressures that the nuclear reactions such as we have described for our Sun (page 40) began. At the same time, the masses left over from the formation of the star also condensed into discrete bodies. As these grew, their gravitational attraction caused more mass to be gathered up until these proto-planets effectively swept clean the space around the star of most of the debris left over from the original nebula.

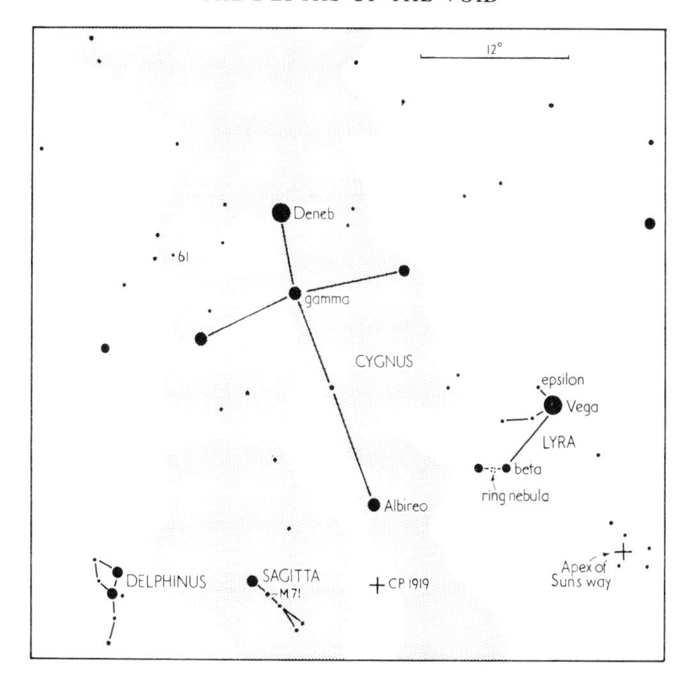

31 Star chart 6. The first pulsar (CP1919) and the best candidate for a black hole, Cygnus XI (very close to Gamma Cygni) are found in the 'northern cross', Cygnus. Also a rift in the Milky Way caused by dark obscuring dust and gas can be seen along the length of the constellation. Albireo (Beta) is a magnificent wide double. The first star for which parallax was measured, 61 Cygni, can be found from Deneb. The Sun is moving through space towards a point near Lyra, a constellation which contains the planetary nebula M57, the Ring Nebula, the type star of the Lyrid variables, Beta, and the famous double double, Epsilon 1 and 2 Lyrae.

While models of star formation such as this have existed for a long time now, there were many unsolved contradictions in the theories put forward when they were worked out in detail. The first problem is that unless a very massive concentration of gas exists, the weak effects of gravity are insufficient to overcome the growing pressures in the gas, and the centrifugal force of the rotating mass. It appears, for example, that the Solar System and other stellar systems, such as the binary stars described earlier, have too much angular momentum to have allowed such condensation.

Theories of star formation were bounded only by mathematical considerations until comparatively recently, when new observational data were found.

Ionised hydrogen can be observed in the emission spectrum of the glowing gases comprising emission nebulae. The most famous of these is perhaps the Great Nebula in Orion's sword (see Plate 12). These glowing nebulae are known as HII regions, and during the 1940s the astronomer Bart J. Bok drew attention to small, dark objects seen in photographs of HII regions such as the Orion nebula, suggesting globules of dust or other obscuring matter seen against the brightness of the nebula. These globules must be light years across, of course, to be visible at distances such as that of the Orion nebula, about 460 parsecs. The term 'globule' seems appropriate, however,

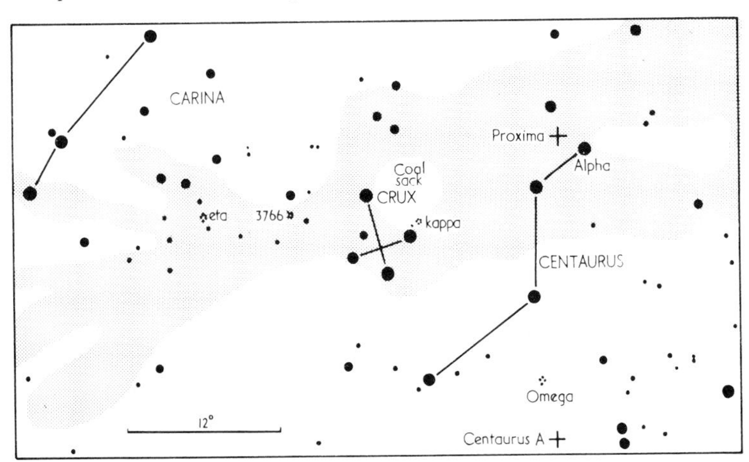

32 Star chart 7. The Southern Cross, Crux, is close to another famous dark nebula in the Milky Way, the Coal Sack, on the edge of which is a magnificent cluster surrounding the red star Kappa. The nearest star, Proxima, is very faint, but nearby Alpha, a beautiful binary, is easy to see in Centaurus. Also in Centaurus is Omega, the finest globular cluster in the skies, and the famous radio source, Centaurus A. The other side of Crux is Carina containing a rich cluster, 3766, and the famous Keyhole nebula surrounding Eta.

because they appear distinctly spherical in shape.

At about the same time the first true molecule was found in interstellar space, the free radical of carbon and hydrogen, CH, detected by absorption lines in the spectra of very bright stars, such as types O or B, indicating the presence of the molecules in the space between the star and Earth. The molecules of ionised CH and the radical CN (cyanide) were found in the same way within a few years. With the advent of radio astronomy, interstellar atomic hydrogen was discovered by the famous emission at 21cm wavelength. This provided a most important new method of probing our own galaxy, the centre of which is completely obscured to optical wavelengths by the interstellar dust and gas. Although this dust is densest at the galactic centre, radio waves can still penetrate.

Radio astronomers then discovered a new molecule, the hydroxyl radical, OH. It was anticipated that this would be found in the HII regions as an absorption line in the spectrum, but in fact the molecules were found in these regions by an intense *emission* of energy, rather than as an absorption. It is believed that this is due to a sort of chain reaction in the hydroxyl in which, as a molecule drops to a lower energy state, it emits energy at the characteristic radio wavelength of its initial state. This energy stimulates other molecules to do the same, causing an intense emission of energy. At the same time, some source of energy within the cloud stimulates the molecules to higher energy levels, so sustaining the process. The effect has been described as an interstellar *maser*, the maser being the radio-frequency equivalent of the optical laser in which light-energy emission is stimulated at characteristic frequencies in a similar manner.

Loss of energy by maser stimulation of the hydroxyl radical provided an important clue in the puzzle of the birth of stars. If the star begins life as an enormous accretion of gas and dust, seen in HII regions as a Bok globule, perhaps some of its rotational energy could be lost by this maser emission.

Subsequent studies of regions such as the Orion nebula showed that other molecules were also present. These include formaldehyde (H_2CO), water (H_2O), carbon monosulphide (CS), hydrogen cyanide (HCN), and carbon monoxide (CO), the latter forming an enormous cloud well beyond the visible regions of the nebula.

It is now widely believed that the Orion nebula is a birth-place for stars. As well as many dark globules, there are many very young stars within the nebula. The cloud of gas, and the stars within it, appear to be collapsing towards the centre of the nebula.

Another tool new to astronomy has been important in the study of the Orion nebula. This is infrared astronomy. Infra-red emission represents temperatures from as low as a few hundred Kelvin, say room temperature, up to about 2000K at which temperature red giants such as class-M Betelgeuse become visible at optical wavelengths. However, even bodies at 250K are visible in the infrared wavelengths of the spectrum, but since air absorbs this radiation so readily, it is impossible to detect through the thickness of Earth's atmosphere. There are certain wavelengths which do penetrate the atmosphere, however, and by means of these infrared 'windows' and by experiments carried on rockets, our knowledge of the infrared sky has grown considerably during the last ten years.

One of the interesting discoveries in the infrared was that gas clouds such as the Orion nebula were very bright at these wave-lengths. The cloud clearly contains a large mass of material at comparatively low temperatures.

The Orion nebula shows an enormous region of infrared emission over a light year across, with concentrations of emission in the region of the multiple-star system in the middle of the nebula known as the Trapezium, evidently caused by the stars themselves heating the surrounding gas and dust. But a second concentration of infrared emission in the Orion nebula has been found which is associated with no known radio or optical source, and yet is more intense than the emission from the cloud round the Trapezium, into which energy is being poured by many young bright stars. The second source is comparatively compact, forming a point source of radiation which suggests a star. Yet if this were a red giant, the complete absorption of its light by the dust and gas in the surrounding nebula would demand an extreme reduction in its light output (a factor of 10^{30}). At the same time, to account for the emission at infrared wavelengths it would have to be many times brighter than any known supergiant. The other explanation, now more widely accepted, is that this source in Orion is a protostar with a surface temperature of about 600K. Other discrete sources have been found in this and other nebulae, which would support the visual evidence of the dark Bok globules. These globules have also been detected in regions of space remote from nebulae, so there is no special connection between such globules and the HII regions.

Emission nebulae are often associated with the formation of clusters of stars. This is probably because in the early stages of evolution of the galaxy, matter began to collect under gravitational effects into local concentrations. These in turn developed concentrated masses from which stars eventually formed. The result was local concentrations of stars. Many of these *galactic clusters* are associated with nebulosity owing to the energy the young stars are pouring into the surrounding gas. A famous example is the naked-eye cluster in Taurus called the Pleiades. Seven or eight stars can be seen in this cluster with a keen unaided eye, but with binoculars, very many stars are revealed, and with photography the centre of the cluster is seen to be glowing with energy.

Many galactic clusters feature in Messier's famous catalogue of nebulous objects, and they come in all sizes from naked-eye objects such as the Pleiades and Praesepe to small faint clusters needing a telescope to be clearly seen. The stars in these clusters tend to be young, white stars of classes O and B.

In complete contrast are the tightly-packed clusters of red giants and other older stars called *globular clusters* because of their approximate shape. The globular clusters can contain as many as half a million stars in a space some 60 parsecs across. These clusters were probably among the first stars to form in the Galaxy, even before gravitational effects and the rotation of the Galaxy had produced the flat spiral shape it now has. Accordingly, the globular clusters have positions well outside the normal galactic plane and are found in a halo round our own Galaxy and other galaxies. Early in their life, as they passed through the Galaxy in their often highly elliptical orbits through space, and as the stars in the clusters evolved, the gas and dust normally found in the interstellar medium were either stripped away by the excursions through the heart of the Galaxy or swept up by the stars themselves owing to the dense packing of the cluster.

As well as the early life of stars, infrared astronomy has made discoveries about their old age.

One of the brightest infrared sources known is the emission nebula surrounding Eta Carinae in the southern sky, about 12° west of the Southern Cross. It is known as the 'Keyhole nebula' because the dark clouds that obscure parts of the bright emission nebula appear to some (with lively imaginations) as a keyhole. The star Eta itself flared to magnitude –1 in 1843, but has since remained at about magnitude 6, best seen with binoculars or a telescope. This suggests that Eta Carinae is heating up the clouds of dust that surround it in the nebula, but in this case the star is more likely to be nearing its end than in the process of being formed.

The evolution of stars

On page 110 the Hertzsprung–Russel diagram shows the relationship between the surface temperature and spectral class of most stars. Red giants and supergiants, and white superdense dwarf stars, clearly fail to fit into the 'main sequence', but by and large the diagram suggests that stars begin life as huge clouds of condensing matter that gradually increase in temperature as their internal pressures rise under the influence of gravity, thus shrinking and increasing in temperature until they become yellow giants, then intensely hot blue stars (about class O) then cooling through yellow dwarf stages (such as our Sun) to red dwarfs, and finally a burned-out cinder. This comparatively placid existence fits the main sequence nicely, but it was soon clear that gravitational energy alone could not sustain the star for more than quite a brief lifetime.

When Hans Bethe and Carl Weizsacker proposed the atomic conversion of hydrogen to helium (see page 41) via the carbon–nitrogen cycle or similar mechanisms, the amount of energy available in a typical star could now be matched to its lifetime, but clearly the sequence of events could not be as straightforward as originally proposed.

Stars, it is now generally believed, evolve through many of the stages represented on the Hertzsprung–Russel diagram, but they do not necessarily all pass along the same path. Many variable factors in the early life of the star cause it to follow quite different evolutionary paths. The original 'mix' of materials in the condensing proto-star, the total mass, the speed of rotation, and the position of the star in the galaxy all play a part in determining its fate, it is believed. However, most stars pass through a broadly similar sequence. From the huge cold cloud of condensing matter, the star builds up sufficient internal pressure under the influence of its own gravitational forces to increase internal temperatures to a point where the conversion of hydrogen to helium begins, and the star begins to radiate energy as a red dwarf of class M or K. As the star begins to radiate energy it builds up its store of helium and reduces its store of hydrogen. But helium is less 'transparent' to energy than hydrogen, and in effect the build-up of helium makes an insulating blanket around the star's interior, causing the internal temperature to rise further and to accelerate the conversion of hydrogen to helium. The surprising result is that the energy produced by a star continues to increase despite the fact that the hydrogen fuel is being used up. By the time the star has passed through classes G and F the store of hydrogen is getting noticeably reduced. In addition, the nuclear processes that accomplish the conversion of hydrogen to helium are likely to change from the proton–proton reaction at lower temperatures to the carbon–nitrogen cycle which uses up hydrogen more quickly than ever.

The stars of class G, such as our Sun, are therefore about half-way through their life as 'ordinary' stars, and are at a fairly stable stage during which hydrogen is being used up at a steady rate. Later in life (only between 1000 and 10 000 million years hence!) the Sun will be using up almost the last of its store of hydrogen and will have become so hot that life on Earth will have become completely impossible.

When the internal temperatures of a star reach these levels, the outer regions begin to expand and cool. If the star is big enough, it could then expand to enormous dimensions, when it would be classed as a supergiant such as Betelgeuse. The Sun, for example, should become only a modest red giant. In either case, the star is now being powered by a dying core of degenerate matter. The energy output can vary as the interior of the star becomes rather poorly mixed so that reactions take place unevenly, and the variability characteristic of many red giants then takes place.

Finally, as the centre of the star becomes exhausted, it cannot sustain even its tenuous outer regions and it begins to shrink once more under gravity, its temperature rising due to the increasing pressure alone. But this time, the star is composed of heavier elements which cannot be converted like hydrogen to helium, so the nuclear furnaces are virtually dead. Gravitational pressures alone cannot sustain an energy output for long, so the star becomes a white dwarf of incredibly dense degenerate matter such as Sirius B and then cools down to finally die.

But there are several other ways in which a star may evolve. At the red giant stage instead of a little peaceful variability it may undergo a violent internal outburst of energy which throws off the entire outer shell of tenuous matter in a temporary flare-up or explosion, which can be seen across a galaxy and which is called a nova.

The star may suffer an even more violent explosion leaving only a tiny core of degenerate matter behind: material so dense that atoms are stripped of electrons and are packed closer together than is possible in normal matter. The rest of the star's material is flung into space where it can be seen as the remnants of such supernovae hundreds or thousands of years later (which of course is only a short time in astronomy.) Supernovae are rare indeed, the last seen in this galaxy was in 1604, seen by Kepler and his contempories, before telescopes were turned on the sky!

Perhaps the most famous supernova in our Galaxy was the one in Taurus recorded by Chinese astronomers in 1054. This was particularly important because the result of that colossal explosion can be seen today. With a modest amateur telescope, very close to the tip of the southern horn of Taurus the Bull (the star marking that southern tip is Zeta Tauri) is a faint patch of light. In photographs taken with huge telescopes, it can be seen that this patch of light is a fantastically patterned cloud of glowing gas. This nebula was the first to be listed in Messier's catalogue of nebulae, galaxies and clusters and was called the Crab nebula after a detailed drawing made by the Earl of Rosse which showed its filamentary nature for the first time. If the first photographs taken of the Crab nebula some 60 years ago are compared with modern photographs, it is clearly seen that the cloud of gas is expanding rapidly. In fact, by shrinking the Crab nebula at its known rate of expansion we can work backwards to the date it was formed, 1054, which is one of the proofs that the supernova seen by the Chinese at that date was the origin of the nebula. The Crab has been a constant source of important data and discoveries, and we shall return to it shortly. But first we must conclude our brief review of stellar evolution.

The transition from red giant to white dwarf, or other superdense star can take place cataclysmically, as described above, or over a longer period of time. It may also take place via a completely different sequence. The hottest stars, such as class O and Wolf-Rayet (class W) stars appear to start life from something like a red giant, with a dense high-energy core and a very large outer atmosphere emiting ultraviolet radiation. The spectra of these stars include bright emission lines similar to the bright emission nebulae so they probably resemble a gaseous nebula with a central star pouring energy into it. These stars then shrink and enter the top end of the main sequence, rapidly degenerating to another giant stage and shrinking to a blue-white dwarf. The entire lifetime is probably only 10 per cent the length of more common main sequence stars.

The American astronomer Harlow Shapley noticed from studies of globular clusters that they contained no blue supergiants or Cepheid variables and, as explained earlier, he noted the remarkable absence of cosmic dust and gas. The most luminous stars found in these clusters were red giants. A similar type of distribution of stars was found by Baade at the nucleus of the spiral galaxies, in the elliptical galaxies (see next chapter), and even in the space between the spiral arms of the galaxies. These became known as Population II stars, as distinct from most of the stars in the spiral arms of the galaxies in general, which are called Population I stars. A roughly spherical halo of Population II stars also surrounds many galaxies. These locations all point to the greater age of Population II stars. It was in the course of studying these stellar populations that Baade concluded that the period/luminosity relationship of the Cepheid variables that had been used for determining the distance of very distant objects such as the galaxies, was in error, a Population I variable being much more intrinsically bright than the older Population II variable. Since the Cepheid stars were brighter than had been supposed, it followed that they were further away than had been assumed.

But that sort of discovery is nothing new in astronomy. The scale of the Universe has generally been underestimated since Ptolemy, and the surprises it can spring have become stranger each time.

Signals from another star

In 1967 one of the team of Cambridge radio astronomers, Jocelyn Bell, found that during the course of one experiment a signal had been picked up which repeated rapidly, by astronomical standards, every 1.3 seconds. Moreover, the signal repeated with amazing accuracy. So odd was the signal that the possibility of an alien artificial source was seriously considered for a while. Before long, other sources of a similar type had been found, including one at the centre of the Crab Nebula, and it was suggested that these strange objects might be related to the postulated 'neutron' star, the next stage in the degeneration and gravitational collapse of a white dwarf.

If the mass of the white dwarf were sufficient, it had been argued, gravitational collapse would continue to pack the very nucleus of each atom next to its neighbour, producing an incredible mass in a very small volume. Now gravity is a weak force compared with other forces, but when matter is packed as tightly as in a neutron star it becomes a very powerful force indeed, and could result in some very peculiar phenomena. The rapidly varying signals could originate in pulsating stars, which were soon called *pulsars* for short. These are now generally thought to be neutron stars rotating on their axis once every second or so. The radio pulse would then be something like a lighthouse beam being sent out from the star, although the mechanism which could cause this was not really understood when the suggestion was first put forward.

The smallest neutron star might have a mass about $1\frac{1}{2}$ times that of the Sun, but its diameter would be only about 20km! Conservation of angular momentum would cause the star to speed up as its diameter shrank—although this can sometimes be a relatively sudden process since it appears pulsars can also result from supernovae. In addition, the neutron star would have a very-high-density magnetic field, it is thought, since the bulk of the star would be effectively a neutron 'liquid', but there would be enough charged particles (electrons and protons) left in this liquid to make it a very good electrical conductor. Thus the neutron star is a giant celestial dynamo producing energy in the form of intense magnetic fields which generate the signal picked up on Earth. Where is the energy coming from?

It was soon found out that the rate of the pulses from the pulsars was steadily decreasing, and at quite a high rate by astronomical standards. Since the Crab nebula pulsar was formed, for example, its rate of pulsing has halved. This neutron star is now rotating 30 times a second and producing some 10^{30} watts of power which is exciting the familiar glow in the surrounding nebula.

Using a special image-intensifier technique to produce a visible image from the optically very faint central star of the Crab nebula, it was found that the star pulses in the visible spectrum at the same frequency as in the radio spectrum. Now it is not easy to tie in an explanation of the optical pulsing with the widely accepted model of the rotating magnetic field.

One popular alternative is to consider the magnetic field at a distance from the axis of rotation at which the magnetic field is being swept round with the star at nearly the speed of light. (Rather like the end of a piece of rope that you hold while spinning round on the spot fairly slowly: if the rope is long enough, the velocity at the end would eventually approach the speed of light. With the incredible speed of the pulsar's rotation, this is reached quite close to the star.)

There is thus a 'cylinder' around the axis of the pulsar at which the magnetic field is sweeping around at the speed of light. Some source of particles close to this speed-of-light cylinder would emit radiation of the type observed (covering a wide spectrum) but it is not clear what the nature of such a source would be. It has been proposed that bunches of particles near the speed-of-light cylinder could cause cyclotron radiation. Maser emission could be stimulated in a similar way. The mystery remains, and is likely to do so for some time to come. Neutron stars were suggested before pulsars were discovered, and it now seems fairly likely that neutron stars are in fact the cause of the pulsar phenomena.

The oddest thing in the Universe?

An even odder object had been suggested by the theoreticians, and although the evidence is still tenuous, it seems increasingly likely that these too may exist in large numbers:

the only problem is that for all intents and purposes they are not there! Consider the neutron star, which, if it were possible to remove anything from the star, away from the intense gravitational field necessary to hold the material together, would contain thousands of tonnes in each piece the size of a pinhead. Could it collapse still further, if it were sufficiently massive to start with? Could the neutron star itself collapse still further when its rapidly radiating rotational energy is dissipated? Eventually, the intensity of the graviational field would be such that not even light or any form of radiation could escape the gravitational pull of the object.

Einstein predicted that light would be distorted by a strong gravitational field, and this was found to be true when it was observed that stars whose position was known with great accuracy had apparently been moved by the proximity of the Sun's limb when observed during a total eclipse of the Sun. The reason for this, Einstein had said, is that gravity is a distortion of space itself, so that the presence of the Sun means that the space between us and the star is slightly 'bent'.

In the case of a massive neutron star, if the gravitational field was intense enough, the space around the star would be folded in upon itself completely. The star would be in a position analogous to that of an ant standing on the surface of a flat soap film which slowly bends under its weight until the film forms a bubble completely around it, then drops away from the film with the ant inside the bubble. To another ant on the surface of the soap film it seems as if his friend has left the Universe altogether. Of course, such analogies are full of misleading ideas and must not be taken too seriously. But the net result of our super-collapsed star is that it would disappear; no light would emerge from it, or be able to pass through the gravitational field it produces. Thus such objects have acquired the name *black hole*.

A lot has been written about 'black holes' by now, so that many people may assume they actually exist: well, they may, but so far there is no proof that they really do exist.

If the object cannot be seen, it is reasonable to ask, how can it be detected? The enormous gravitational field of the black hole would extend well beyond the space it occupies.

There are a few locations where there is evidence of something peculiar which *may* be a black hole. In each case something which is clearly *not* a black hole is in orbit around something which *might* be. In such a strange binary star, gravitational forces would be progressively disrupting the object in normal space, and the black hole would be dragging everything that came near enough into itself like a giant celestial plug hole. Like the plug hole, matter orbiting the black hole would travel faster and faster as it approached the point where it would finally disappear from the Universe as we know it.

In theory, the behaviour of a black hole is simple, compared with some problems in physics, and the case of the binary star incorporating a black hole provided astronomers with the best chance of finding observational evidence of their existence. If an ordinary star was pouring its material into a neighbouring black hole, the dust and gas would go into orbit around the black hole and as friction between the gas and dust developed it would cause it to spiral inwards towards the black hole and at the same time its temperature, already high from the parent star, would climb from some 10^5K (one hundred thousand Kelvin) to an incredible 10^8 or 10^9K, as the matter neared the black hole itself. This high energy would result in the emission of X-rays, and so to find possible candidates for black holes the astronomers needed to find a compact source of X-rays, associated with a spectroscopic binary star. Such a star was known: Cygnus X-1.

The X-1 simply means the first object in a catalogue of X-ray objects. These were first detected by rocket and balloon-borne experiments that could carry detectors above the thickest parts of the Earth's atmosphere which acts so effectively in screening the surface of our planet from X-rays. Early in the history of X-ray astronomy (which began in about 1963) it was evident that most of the X-ray sources were in the region of the Milky Way, so that they were distant objects in our galaxy. Cygnus X-1 is in the middle of the Milky Way in the northern part of the constellation of Cygnus, the 'Northern Cross'.

X-ray astronomy started to make great advances with the launching of the US–Italian satellite *Uhuru* in 1970 which enabled X-ray sources to be discovered and examined in much greater detail than had ever been possible before. This revealed for the first time that X-rays emitted by the source Centaurus X-2 varied with a precision similar to the pulsars with a period of 4.84 seconds. Then another, Hercules X-1, was discovered to have a period of 1.2 seconds. When the variations in X-ray emission from Centaurus X-3 were studied in greater detail, it was found that the period varied by some 0.001 per cent, which may not seem much but is in fact a thousand times less accurate than a normal pulsar. The magnitude of the X-ray emission also varied by a factor of about ten, and it was found after some months of watching these two variations that they repeated themselves over a cycle of exactly 2 days 2 hours $5\frac{1}{4}$ minutes. This immediately suggested a pattern such as the light curve from Algol, in other words that the source was a close binary system. It was not till 1973 that a Polish astronomer, Krzemiński, found a very faint star with a light curve corresponding to the X-ray variations of Centaurus X-3 at about the right position in the sky. It was examined with the spectroscope and given an intrinsic luminosity of 100 000 times that of the Sun. Its faintness is due to its being in the central plane of the galaxy where obscuring matter such as dust and gas severely reduces the brightness of the object, and its distance was estimated as 7600 parsecs.

The clue to the nature of bright X-ray sources had been found. Was Cygnus X-1, one of the brightest X-ray sources in the whole sky, also a binary? Radio emissions were discovered from the same area in 1971 and a supergiant similar to Centaurus X-3 was found to be at the same point in the sky as the radio source. No radio source had been found previously at that position. The X-ray emission was found to have doubled in average intensity at the same time, giving almost conclusive evidence that the radio source, the supergiant and Cygnus X-1 were the same object. It was also found that the spectrum of the supergiant indicated that it was a binary, with a period of 5 days 14½ hours. More recent observations made at the Kitt Peak Observatory indicated that the visible star had a companion with a mass of about 8.5 times the mass of the Sun. If there are any other component stars in this system they are insufficiently large to disturb the orbits of the two main bodies, so that the invisible companion seems almost certain to be greater than the mass needed to ensure a black hole.

Inside a black hole

What do astronomers think a black hole is like? From ordinary physical laws such a Newton's we know that if a planet were compressed to a quarter of its original diameter, a body on its surface would be four times closer to its centre and, as the total mass has not changed, the energy needed to escape the gravitational pull would be four times what was needed from the original surface. A rocket fired from the surface of a planet must reach a certain velocity to go into orbit, and must reach a speed called the *escape velocity* if it is to escape from the planet entirely (assuming that it is required to fire the rockets once and coast away from the planet).

The escape energy is proportional to the square of the escape velocity, so by compressing the planet to one quarter of its original size we double the escape velocity. If we further reduce the size to one quarter again, we get a further doubling of the escape velocity and so on.

A body the mass and size of the Earth has an escape velocity of 11km/s. By the time we have compressed the Earth into an object some 10mm radius, the escape velocity is the speed of light. This is a critical size, called the *Schwarzschild radius* after the German mathematician who studied this effect. The behaviour of matter then may become much more complex than Newtonian physics allows us to contemplate. Nothing happening inside that critical radius can be known to us, since light and all electromagnetic radiation is confined within the black hole. This has therefore become known as an *event horizon*. If the body were to continue to collapse in a perfectly radial direction, a point of infinite density would be reached where space and time had no meaning as we know it: time would literally stand still. This is known as a *singularity* in space–time, which means basically that normal space–time is interrupted. The vertex of a pyramid is, in effect, a singularity in space, since anywhere away from this point you know what plane you are climbing on, but at the point itself you are not on any particular side of the pyramid.

So the centre of the black hole is a singularity and, at the critical radius from this centre, the event horizon marks the outer edge of the black hole. This is in turn surrounded by any matter in the vicinity, in a very close and rapid orbit around the event horizon compressed into a flattened disc by the combined effects of kinetic energy and gravity. This disc, called the *accretion* disc, would contain a great deal of matter at very high temperatures in the case of a binary black hole.

But black holes may be stranger yet. The astronomical world hailed a new and complex theory by Stephen Hawking, a mathematician from Cambridge University, when it was published in 1974. Hawking suggested that radiation *could* escape from the region of a black hole. The basic idea was that as the black hole accumulated matter, it would get bigger—or rather the event horizon would get bigger. Depending upon the surrounding space there could come a time when the black hole could lose energy to its surroundings, possibly cataclysmically in an explosion which would cause the disappearance not only of the black hole, but also the matter which had formed it in the first place, the final dissipation of the energy locked up in the mass that had degenerated inside the black hole. Hawking showed that there would be a definite time for such an evolution to take place, and that a massive black hole such as Cygnus X-1 would be so big that it would not finally explode for a period vastly longer than the present age of the Universe. But it has been suggested that when the Universe was formed, according to the mechanisms now believed to have been involved, many lumps of matter no larger than a mountain might have been compressed to their critical radii by the tremendous forces at work, whereupon they would collapse to singularities inside an event horizon about the size of an atom. The lifetime of these mini-black-holes would be comparable with the presently estimated lifetime of the galaxy. The theory could therefore be checked if we could detect the final explosions of some of these mini-black-holes, since we may have to wait 10^{66} years or so for something like Cygnus X-1 to pop off. One way this might be done is by detecting discrete bursts of gamma rays. But like X-ray astronomy, gamma-ray astronomy is still in its infancy and no detectors sensitive enough have yet been built.

Black holes have provided the mathematicians with some of the most intriguing possibilities for studying the implications and checking the validity of Einstein's general theory of relativity, and possibly unifying quantum physics and relativity. Black holes may be at the centres of the mysterious quasars (see next chapter), or they may be the ultimate beginning and end of everything in this Universe. So there it remains.

4 The Universe of Islands

Charles Messier is remembered for his catalogue of nebulae and star clusters, begun in August of 1758 when searching for a comet he had found some weeks earlier and lost again. He found a hazy patch which might well have been the comet near the star Zeta Tauri. After observing the object for a while, it became evident that it was not moving against the background of stars, so could not be a comet: Messier was looking at the Crab Nebula! Messier decided to search out all these hazy, comet-like objects in the sky so that they would not confuse his comet hunting. He listed the Crab Nebula (although it acquired the name much later) as the first in his new catalogue, M1. He described it as a 'nebulosity above the southern horn of Taurus. It contains no star.'

The second object in Messier's catalogue, M2, was a globular cluster in Aquarius. In a moderately good quality amateur telescope, this appears as a compact ball of faint stars, but to Messier it was a 'nebula without star'. (Neither M1 nor M2 were discovered by Messier, but the third object—another globular cluster, in Canes Venatici, easily resolvable into a fantastic globe of faint stars in an ordinary amateur telescope —was the first of the 17 new discoveries Messier was to make in his catalogue of 104 objects.)

Apart from the Crab Nebula and a few gaseous nebulae, the first 30 objects were either globular or galactic clusters. Messier knew that some 'nebulous' objects could be resolved into clusters of stars, stars within our own galaxy which are clustered together like the Pleiades, a naked-eye cluster, or Praesepe in the centre of Cancer. M31 was almost one-third of the way through the catalogue and yet had been known to the ancient astronomers who could see a faint patch of light, also very nebulous in appearance, in Andromeda.

Messier was quite used to 'nebulae without star', but although he was naturally impressed by the Great Nebula of

Andromeda, he failed to recognise its real nature.

It was not until about 50 years later that William Herschel, with the advantage of his superior instruments, postulated that some of these nebulous objects (which he also catalogued, and to a much greater extent than Messier) might be outside our Galaxy altogether. When the huge telescopes of the Earl of Rosse in the middle of the 19th century detected the spiral structure of some of the Messier objects, including M31, the term 'spiral nebula' was coined, but the general opinion was still that like the resolvable clusters, the spiral nebulae were situated inside our Galaxy.

It was not until 1924 that Edwin Hubble detected Cepheid variables in the Andromeda spiral, and deduced that it was about 300 kiloparsecs distant, well outside the known confines of our Galaxy. In fact the Andromeda galaxy is over twice this distance away, so is the most distant object that can be seen with the naked eye.

To find M31, it is best to choose a clear moonless night in autumn, and locate the Great Square of Pegasus, in the south at midnight in mid-September. The northeast corner of the square is Alpha Andromedae. If you include this star, Andromeda contains three fairly bright stars and a fourth, less bright star in a line curving to the north, beneath the W of Cassiopeia. These are the main stars of the constellation Andromeda. The star next to the eastern end is Beta Andromedae, and from it northwards towards Cassiopeia is a line with two other dimmer, but easily visible stars. Just to the northwest of the end of this line you will see the faint patch of the Andromeda nebula. In a small telescope or binoculars it can be seen clearly, but it is still only a nebulosity showing no detail. In fact, what you are looking at is only the central regions of this galaxy. The whole object, including the fainter outer regions, is over six times the apparent diameter of the Moon in length and twice in width. From its distance, 680 kiloparsecs, and full extent, about $3\frac{1}{4}°$, its diameter can be simply calculated: $680 \times \sin 3.25 = 38.5$ kpc. In fact, the full extent of the galaxy is some 39 kpc, compared with the diameter of our own Galaxy, 25 kpc.

The distance of the Andromeda galaxy was not determined accurately (assuming that we have got it right now) until as recently as 1952 when Baade concluded that the Andromeda Cepheid variables used to determine its distance were Population I types, and therefore very much more intrinsically bright than had been thought.

Once extragalactic objects had been shown to exist, the term 'nebula' became confined to gaseous emission nebulae or absorption nebulae, that is, clouds of gas, while other objects are known as clusters (globular or galactic), planetary nebulae (clouds of gas surrounding some stars which may have undergone cataclysmic changes such as shrinking to white dwarfs, or having been a nova), or galaxies.

As well as investigating the periods of Cepheids in the Andromeda nebula, Edwin Hubble turned his attention to the galaxies in general. A well-known grouping of galaxies occurs in the sky in the constellations of Comae Berinices and Virgo. Here, in the region of the sky north of Spica in the heart of the constellation Virgo, is an apparently empty sky. But on a clear dark night, this area glistens with faint light from thousands of faint stars and even fainter galaxies. Hubble studied these galaxies in 1926, classifying 167 in this group into either spiral, elliptical or irregular galaxies. Later, two distinct types of spiral galaxy were identified, and this was refined further in more recent years to give sub-divisions of these and the other types in a way thought to represent their possible evolution.

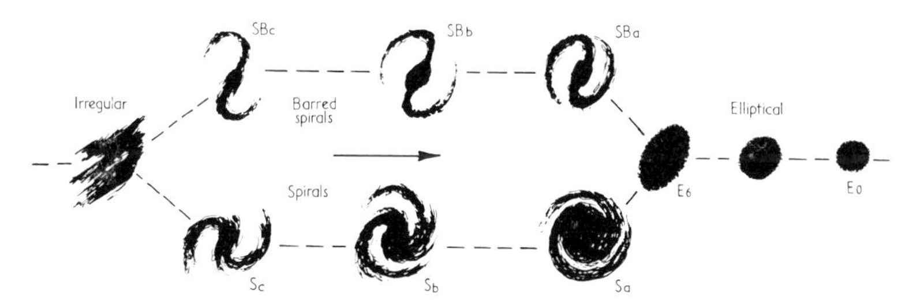

33 The classification of galaxies originally proposed by Hubble. But it is now thought that galaxies evolve from an irregular form to tightly packed ellipticals (as indicated by the arrow).

Hubble also examined the spectra of distant galaxies. The distances of these had been inferred from their apparent magnitude compared with closer galaxies in which Cepheid variables had been found and measured. It was also known, from earlier work by V. M. Slipher, that the features of the spectra of distant galaxies were displaced towards the red end of the spectrum, this *red-shift* resulting from a lengthening of all the optical wavelengths thought to be due to the Doppler effect, which meant that the galaxies in question must be receding from our own galaxy at enormous speeds. The amount of red shift indicates the velocity with which the galaxy is receding.

Hubble found that most galaxies were receding from our own, except for the Andromeda galaxy, the Greater and Lesser Magellanic Clouds, and a few others which appeared to be grouped together in space and moving through the Universe more or less together. Not only were the galaxies apparently rushing apart from each other, except for the 'local' group, but the further away they were, he found, the greater the speed of recession.

This had two sensational implications: firstly, galaxies throughout the visible Universe were accelerating apart from each other. They can be likened to spots on the surface of a balloon which is being blown up, inside which is another balloon covered in spots, and another inside and so on, the balloons being inflated faster the nearer they are to the outside. Moreover, the most distant galaxies, and hence the fastest receding, were being photographed in Earth-based observatories millions of years after the light being recorded left the galaxies in question. The further astronomers looked into the distance, the further they were looking into the past history of the Universe. Optical telescopes could record images of galaxies over 200 million parsecs (200Mpc) distant, but when the famous Mount Palomar 200-inch telescope was first turned on the skies after the World War II even these distances shrank as more and more distant galaxies were discovered.

But Hubble, and his colleague Milton Humason, working with the 100-inch telescope of Mount Wilson Observatory, had found a second vital implication: if the galaxies were accelerating apart, by working backwards from the observed state of the Universe, a time would be found when all the galaxies were lumped together in what came to be called the primeval 'atom', meaning an almost infinitely dense mass containing all the atomic particles in the Universe today in a ball only a few light years across. (In fact, the description suits a large black hole which, as has already been suggested, may explode after an enormous time.)

Hubble brought all his observations together to form a possible evolutionary theory of the Universe in which a cataclysmic event started the expansion of the Universe, which he estimated took place 2000 million years ago. Modern theories put the beginning of the Universe at a much earlier date, possibly as long as 25 000 million years ago. Many of the 'atoms' from that initial 'big bang' (the most classic understatement in astronomy) condensed gravitationally into lumps of matter that became the galaxies. Hubble supposed that matter was attracted into spherical volumes that became spherical galaxies, which then developed spiral form, the arms of the spiral getting more loosely wound as the galaxies aged until the galaxies became irregular in form. Present theories tend to favour the evolution of galaxies in the opposite direction, so that they end life as small, densely packed objects resembling the nucleus of a spiral galaxy.

There were many astronomers who found unacceptable the idea of a Universe having a definite beginning, and an end which appeared to be a final separation of the galaxies at speeds greater than the speed of light (in other words, when neighbouring galaxies would disappear for ever). If the Universe began as a big bang, and the galaxies were accelerating apart ever since, they argued, then the most distant reaches of the Universe would be populated at a very different density to the nearer galaxies. But all observational evidence indicated a fairly uniform distribution of the galaxies throughout space. But the galaxies were undoubtedly rushing apart, so unless there was an entirely different explanation of the red shift (in which case the entire fabric of modern cosmology and cosmogony would be destroyed), that meant that matter was still being created to fill the voids formed between the galaxies.

The rival theories of the formation of the Universe, big bang and continuous creation, caused heated arguments in the astronomical world for many years. Even Palomar's photographs of galaxies receding from our own at 144 000km/s, half the speed of light, failed to resolve the dilemma. The astronomers had to await the development of a completely new observational technique: radio astronomy. With the advent of radio telescopes, new data were rapidly accumulated. In particular, as we shall see shortly, the discovery of a universal background radiation, which appears to be the remnants of the big bang itself, has almost finally disposed of the continuous creation theory.

Probing the ends of the Universe

The first time telescopes were turned on the stars was a turning point in astronomy, as we have seen. The same has applied to the beginning of observational methods using radio, X-rays and the other frequencies of the electromagnetic spectrum.

Electromagnetic energy is radiated when an atom is altered in some way. There are many ways this can be done, but the result is a definite quantity of energy emitted at a particular frequency and wavelength. In physics, these quanta of energy can also be regarded as particles having a particular mass; because, in the end, the picture we have of an atom as a sort of central body with electrons spinning around it like planets is only one of a great many models invented to try and explain the behaviour of the atom.

In 1926 Schrödinger showed that in a free state, electrons behave like particles, and when they interact with other matter they have properties like electromagnetic waves. De Broglie showed that particles of mass m and velocity u have wave properties corresponding to wavelengths $\lambda = h/mu$, where h is a constant.

Energy is therefore emitted at different frequencies or wavelengths (it does not matter which we use, since they are related to the speed of light in a vacuum by the simple relationship $f\lambda = c$ where f is the frequency in hertz; λ is in metres and c, the velocity of light, is 3×10^8 m/s).

Energy with wavelengths of between 4×10^{-7} and 8×10^{-7} metres excites the human eye and is termed 'light'. Just beyond the visible spectrum is infrared radiation which we experience as heat, and this has longer wavelengths of up to about 2×10^{-4}m. Beyond infrared is the spectrum of radio waves and beyond the longest of these are the low-frequency wavelengths that are audible to the human ear (although this is no longer just electromagnetic energy, since sound needs a medium, such as air, to be transmitted).

At the other end of the visible spectrum is the ultraviolet and X-ray spectrum which, like blue light, is mostly filtered from the light reaching the Earth by our planet's thick atmosphere.

With the advent of rocket-borne astronomical experiments, and more important still, with the advent of astronomy from artificial satellites, the wavelengths previously denied astronomers have become accessible. There is now an enormous amount of data to examine and the birth of X-ray, gamma-ray, and infrared astronomy is similar to the birth of optical astronomy. Most of the sources of energy in the Universe emit across the spectrum and many are far 'brighter' at one of these invisible ends of the spectrum than they are in terms of light output. We have just seen how Cygnus X-1 is very faint, yet one of the most energetic X-ray sources in the sky.

This explosion of information at other frequencies began with the most important modern addition to the astronomers' armoury of techniques, radio telescopes. The radio emissions from the stars were discovered in 1933, as so much has been in astronomy, by accident. An American radio technician, Karl G. Jansky, built a steerable radio antenna as part of his efforts to investigate the cause of static in short-wave radio receivers. He found a steady signal in the decametre waveband that increased in intensity with the position of his steerable aerial, or according to the time of day. He found that in fact the radio emission was from the Milky Way, and moreover increased to a maximum in Sagittarius, which is the constellation where the centre of our Galaxy is situated.

Jansky was not an astronomer, and failed to follow up his momentous discovery. But a keen amateur American astronomer, Grote Reber, followed up this work with a steerable aerial of his own, in 1937. This was dish-shaped, bringing incident radio energy to a focus where the receiver itself was situated, similar in many respects to the modern steerable-dish-type radio telescope. Reber found several discrete sources of radio waves besides Jansky's signals from the Milky Way, and he also found that there were no particularly bright stars in the positions where he found the radio sources.

Radio astronomy was interrupted by the war, but the work on radar during the war was a great boon to radio astronomy. After the war, giant radio telescopes began to be built in various shapes and sizes, and within a few years had transformed the whole subject. New kinds of objects such as pulsars, and radio galaxies were found, together with a new kind of object that defied imagination: quasars.

As part of the routine follow-up to a survey of the sky at radio frequencies, the optical astronomers try to identify the radio objects with something that can be photographed or seen at optical wavelengths. One of the objects in a catalogue compiled by the Cambridge radio observatory was identified as a blue star with a peculiar spectrum. A similar object was identified later at Palomar. The stars in question had enormous red shifts in the spectrum, suggesting they were tremendous distances outside our Galaxy. But the brightness of the objects far exceeded anything possible from a star or even a galaxy. These objects were called quasi-stellar objects, or quasars.

One of the first quasars, 3C 273 (object no. 273 in the third Cambridge catalogue of bright radio sources) has since been the subject of close observation, when it was found to be variable. It has also been observed at infrared, ultraviolet, X-ray and gamma-ray frequencies as well as at radio and optical wavelengths. It is the brightest of all the quasars discovered, some 100 times the total brightness of our Galaxy. It can vary in brightness at different frequencies with considerable rapidity at times, and produces polarised radiation, also varying. Soviet astronomers found recently that 3C 273 can vary in radio emission by as much as 30 per cent in a few hours, and that when the radio emission is at a minimum, the light from the quasar suddenly becomes polarised.

The behaviour of quasars has still to be explained. Several suggestions have been made, including that they are explosions in the middle of very dense star clusters, black holes in the middle of a cluster swallowing matter at a very high rate, or that they may be very massive stars, rotating too fast to allow the collapse of the star to a black hole, and having an intense magnetic field. So it could be similar to a pulsar, possibly over a million times as massive as the Sun, acting like an enormous dynamo and possibly even flinging gas and other matter out of the object at tremendous speeds into the stars surrounding it. At least three quasars have now been found with jets of gas extending several parsecs from the nucleus each sharing the same red-shift as the quasar itself, and so clearly linked with it. At present the quasars are the most energetic objects known. Quasar 3C 279 for example has been known to flare to an absolute magnitude of about –30, which is 10 000 times brighter than the entire Milky Way Galaxy. At this brightness, it is emitting energy at a rate of 10^{40} J/s (10^{37}kW).

As well as the brightest objects in the Universe, radio astronomy has also revealed what seem to be the largest objects in the Universe. One radio source, 3C 236, appears to be two clouds one each side of a galaxy giving out energy from a volume of space $5\frac{1}{2}$ Mpc across, and the central galaxy is a very 'bright' radio emitter. This object is thus almost ten times as wide as the distance from our Galaxy to the Andromeda galaxy.

A major advance was made possible with radio astronomy simply because it freed observation from the wavelengths that were for ever blocked by the dust and gases that obscure so many important parts of space. In particular, it provides the only method of studying the centre of our own Galaxy. Radio astronomy has confirmed that, as Herschel suspected, and as so many other galaxies are, our own galactic system is roughly lens-shaped, with a swelling in the middle of the disc. But Herschel did not realise that the Sun is situated well away from the centre of the Galaxy. From Earth, we see the Milky Way apparently stretching across the sky all round the Earth, but in fact it is much denser in the region of Sagittarius where the centre of the Galaxy appears in our skies.

We now think that the Galaxy is about 25 000pc in diameter and about 4000pc thick at its centre, with the Sun 8200pc from the centre. The Sun is also about 8pc north of the central plane of the Galaxy. The entire system rotates about its centre, taking the Sun round with it, of course, and taking about 220 million years to make one rotation. The total mass of the Galaxy is estimated at 1.1×10^{11} solar masses, and the total number of stars is about 10^{11}, so that the Sun's mass is about average (although, as already mentioned, the masses of stars do not vary to anything like the same extent as their other characteristics).

The majority of galaxies have two arms, twisted into the spiral or barred spiral pattern. From the Sun's position in the far outskirts of the Galaxy we can observe only the arm in which the Sun is situated, the opposite side and the centre of the Galaxy being obscured by dust and gas. But portions of other arms may be curved round the centre of the Galaxy into observable positions, if there was some way of fitting the stars into a pattern.

The arms of a galaxy contain neutral hydrogen, and the Population I star type. A radio astronomer, A. P. Henderson, recently suggested that the observed positions of the arms, and the positions and proper motions of the hydrogen, did not fit a two-arm structure, and he suggested that the relatively rare four-arm structure fitted the observations of our Galaxy more closely. The Sun also appeared to be about half-way between the two outermost arms in the area of the Galaxy where it is situated. The outermost reaches are then represented by the spiral arm seen as the Milky Way in Perseus.

In the opposite direction, towards the centre of the Galaxy, the inner portions of the spiral arms and the increasingly dense clouds of gas hide the centre from view. But here too radio telescopes have been able to penetrate. Once again, the findings have been dramatic. In a very complex experiment involving the syncronising of results from radio telescopes across the entire United States, astronomers at the US National Radio Astronomy Observatory were able to achieve a resolution far finer than could have been obtained with any one radio telescope. They found that near the centre are several concentrations of radio emission, including one very intense source less than one thousandth of a second of arc in diameter. This would make the region smaller than the Solar System. Could this be a black hole? Could it be a similar mechanism to that which results in quasars? These have all been suggested, and the picture has a certain unity: but why are the quasars all so far away, so that we see them as they were millions of years ago? This would suggest the birth of galaxies rather than the end. Might the galaxies explode out of black holes, or even, as has been seriously suggested, emerge from another Universe altogether from which it has disappeared in a black hole, emerging in this universe from a 'white hole'?

Speculation has kept well ahead of the observed data, and as long as it does, almost any idea is as good as another! But there is now a very good indication of something unusual going on at the heart of our own Galaxy.

Other galaxies have shown concentrations of mass at their centres beyond what appear to be normal densities. Recent observations of the giant elliptical (Eo) galaxy M87 in the Virgo cluster suggest that it may have a massive object at the centre. This galaxy has a curious jet emerging from it, so that it resembles quasar 3C 273 (see plate 16) in photographs. The jet from M87 is some 1500 parsecs long and it tends to bridge the space between it and galaxy M84 (a spiral type So) in the cluster. M87 is a strong radio source at 18.3 MHz, and an even stronger X-ray source. Its size is estimated as about 39-kpc diameter. The recent discovery is that not only is the galaxy radiating more energy than can be simply explained, but the nucleus appears to be rotating much faster than the surroundings. This could be explained by a supermassive black hole, and provides another suggestion that perhaps elliptical galaxies and quasars have an evolutionary link, and are powered by massive black holes at their centre.

Families of galaxies

Radio astronomy has also shown that the galaxies are much bigger than previously thought, if their entire sphere of influence is included. For example, galaxies are surrounded with an enormous halo of gas. Faint optical haloes have now been discovered at some distance from the main body of the galaxies, suggesting some form of energy being emitted. It is possible that this radiation is from a faint halo of Population II stars. If so, the night sky must be very dark indeed from a planet of one of these lonely stars.

Observation of the galaxies has now been able to show that they group together in clusters, sharing a more or less common motion through the Universe relative to other clusters, from which they are receding. The clusters of galaxies even appear to be grouped in 'super-clusters', which, although expanding, are not receding from the other clusters in the super-group uniformly.

Our own 'Local Group' contains the Andromeda galaxy, M31, the Magellanic Clouds, which are irregular galaxies and satellites of our own Galaxy, another large spiral, M33 (a rather faint but fine spiral galaxy half-way between the star marking the vertex of the triangle called Triangulum (south of Andromeda) and Beta Andromedae itself from which you begin the search for M31), and many smaller systems. Another dwarf member of the local group, thought to have been separated from the Megallanic Cloud by the gravitational effects of our own galaxy, was discovered in 1977 at a distance of about 150kpc (compared with nearly four times this distance to M31).

A similar group of galaxies to our own local cluster, centred some $2\frac{1}{2}$Mpc from our Galaxy, contains two galaxies close together in the sky in Ursa Major. M81, the brightest of these, is a little bigger than our Galaxy, but contains twice as many stars, making it one of the densest spiral galaxies discovered. M81 can be seen easily with a moderate amateur telescope, and, using a low power on the instrument, a second galaxy, M82, which is irregular, can be seen in the same field of view. It has been said, with much justification, that where you find one galaxy you will usually find another.

Clusters like the Local Group are linked gravitationally. The supercluster to which our Local Group belongs includes the huge group of galaxies in Virgo, centred some 152Mpc years away, and our Galaxy is found on the edge of this supergroup.

Astronomers now believe that the supergroups form supergroup clusters, and so on through a vast hierarchy until one is looking at the whole Universe, when one would find that overall the galaxies are uniformly distributed, but that the Universe is becoming less densely packed as time passes. Several surveys have now been made, simply counting the numbers of galaxies of a given magnitude. One survey, by Shane and Wirtanen of the Lick Observatory, covered most of the northern hemisphere and included galaxies down to 19th magnitude. The average distance of objects in this survey is 420Mpc, and it included about a million galaxies. Such a task is very daunting, and to cover even fainter objects would need a very patient group of astronomers. One experiment has been carried out at the University in Cracow in which a 6° square was surveyed down to magnitude 20.5. This doubles the average distance of the galaxies being counted, and some 10 000 could be found in this small section of sky alone.

The result of these investigations and some extensive analysis has been the discovery that galaxies form quite 'small' dense local groups which are in turn grouped in clusters and superclusters until the clusters are about 20Mpc across, when the distribution of such super-super groups becomes more uniform.

Trying to explain this effect has led to another of the many perennial conflicts in astronomy. The initial distribution of all matter at the time of the big bang must have been very much more uniform than the galaxies are now, so that they have bunched together in groups, clusters, superclusters and so on owing to gravitational forces since the big bang started the expansion of the Universe.

Computer models of a Universe with a hierarchy of groups of galaxies as shown by these surveys indicate that to form the supergroups would need a certain total mass within the supergroup to be able to affect the neighbouring supergroup, and so on. This leads to a figure for the total mass of the Universe which, if true, would be sufficient to ensure that gravitational forces would slow down the expansion of the Universe until it began to contract once more, all the galaxies and groups of galaxies rushing together to form another big bang. A cyclic Universe of this sort appeals to many scientists because it suggests an answer to the problem of what immediately preceded the big bang.

However, some other work quite independent of these studies of galaxy grouping has arrived at a total mass of the Universe well below that needed to halt its expansion, and also apparently too low to explain the formation of the supergroups. The question, therefore, is: will the Universe expand for ever? To examine this problem we must return to the work of Hubble.

Edwin Hubble, as we saw earlier, concluded from his study of the red shifts in the spectra of distant galaxies that the expansion of the Universe began 2000 million years ago. This interval, arrived at by extrapolating backwards in time the observed rate of expansion of the observable universe, became known as the *Hubble Time* in commemoration of this work. Since the value of Hubble Time was first given, it has (not surprisingly) been extended several times. While not all astronomers agree on its value, since its determination can be made in different ways, it is generally held now to be between 12 000 million and 25 000 million years, the usual value being quoted as 19 000 million years. This corresponds to a rate of expansion of about 50 kilometres per second per megaparsec. In other words, a galaxy one million parsecs from our own is receding at 50km/s, while one at two million parsecs is receding at 100km/s, and so on. The rate of recession in km/s/Mpc is called the *Hubble constant*.

The largest red shift known belongs to the quasar 0Q 172 which is apparently receding from us at 91 per cent of the speed of light. This puts its distance at some 5400Mpc (18 000 milion light years), and the quasar is thus being seen close to the beginning of the universe's present history.

No matter in what direction astronomers measure the red shift of distant objects, the apparent distance as judged by apparent magnitude or other distance indicator is in the same proportion to the red shift, and hence to the speed of recession. This indicates a very uniform distribution of the galaxies overall (which was also the conclusion of the surveys of superclusters). Another clue fits the same puzzle: a study of microwave background radiation across the whole sky by Arno Penzias and Robert Wilson of the Bell Laboratories in the United States showed that this was virtually constant, varying by less than one part in a thousand. Such radiation, produced by the effects of the galaxies and the gas in and around them, should reflect the distribution of matter in the Universe, so this also suggests a uniform Universe. The radiation is characteristic of a body at a temperature of 2.7K and is thought to be a remnant of the radiation that bathed the entire Universe at the time of the big bang. It is interesting to note that this background was detected in 1965, but had been predicted in the 1940s by George Gamow, then at the George Washington University.

Gamow had considered the first few hours of the Universe, and concluded that if the primeval atom were composed of energy, its temperature would be something like 250 000 000K when about an hour old. By the time the Universe was 200 000 years old the temperature would have dropped to 6000K, comparable with the present temperature of the Sun's surface. By the time the temperature had dropped to 170K, the pressure of the radiant energy left from the big bang would just about balance the density of matter in the expanding Universe, and at this point, about 250 million years after the big bang, matter could start to accumulate under gravitational attraction, which would increasingly dominate the evolution of the Universe as the background radiation 'pressure' declined still further. So first the clouds of gases began to gather, then they contracted to form the galaxies and the stars within the galaxies and, ultimately, the accreted debris surrounding the stars called the planets.

By the present day, Gamow then calculated, the background would have dropped to about 5K, and this is reasonably close to the 2.7K microwave background discovered in 1965.

The result of these discoveries was a gradual domination of the big-bang theory of the creation of the Universe. As radio telescopes probed deeper into space it became clear that the average density of the Universe was changing so that the concept of continuous creation did not accord with observed data. Cosmologists then turned their attention to the problem of whether the Universe will continue to expand, or whether the galaxies might one day be slowed down sufficiently by the combined gravitational attraction of all the others to cause them to start to move towards each other again, finally to disappear in a cosmic collision resulting in another big bang, and so on.

There are many different considerations which may be used to investigate this possibility: it has already been mentioned that the total mass concentration necessary to cause galaxy-group clustering should be sufficient to halt the overall expansion of the Universe, but the average density of the universe as calculated by several independent means appears to be about 10^{-27} kg/m^3 (the same as one kilogram of material in a spherical volume of space measuring $1\frac{1}{4}$ million kilometres across, a volume only a little less than that of the Sun). At this density the escape velocity (the initial velocity a body must be given to escape entirely from the gravitational attraction slowing it down after 'launch') is only about one seventh the value at which the galaxies are actually flying apart.

To discover the average density of the Universe, it is not necessary to study the entire Universe, or even the entire *known* Universe; the study of the local groups and supergroups has shown that any reasonably large volume of the Universe will be much like another, so we can study groups of galaxies receding from each other at reasonable speeds which are comparatively easy to study. All the galaxies in a sphere of space will be attracted towards its centre of gravity, which is near enough its geometrical centre. The rate of a galaxy's motion away from that centre must have been reducing ever since the initial outward impulse was given during the big bang. This would cause a natural decay in its velocity, that is the change would be imperceptible at first, and slowing at an increasing rate—that is, the deceleration steadily increases. This is the same as the speed of a bullet fired upwards from the Earth: for most of the upward trajectory there would be very little observable reduction in speed, but as the kinetic energy of the bullet became exhausted, the bullet appears to slow suddenly, finally stopping at the top of its trajectory. If you had only crude devices for measuring its velocity, which is the position of the astronomer when measuring the speed and distance of galaxies, you could not detect that the bullet was slowing down until it had almost reached the end of its upward flight.

The critical density for a Hubble Time of 19 000 million years is about 5×10^{-27}kg/m^3, about five times denser than the Universe is, apparently. If this is so, the Universe is 'open': it has no boundary in time or space; it will expand for ever, long after the stars within its galaxies have disappeared into dust, or black holes, or whatever their ultimate fate may be. The Universe may then be represented as a space with negative curvature—the concept defies imagination and may only be considered mathematically, although it does have real implications for what we can hope to see at the farthest extremes of the Universe.

If the density of the universe is over five times higher than it appears, then, ultimately, the galaxies will be halted in their outward flight, and the Universe will collapse. This is called a closed universe; or one with positive curvature.

A factor of five is very little room for error when estimating the total density of the universe. There is so much apparently empty space between the galaxies, for example, that there only need to be one or two atoms of hydrogen per cubic metre more than expected between the galaxies, or the groups of galaxies, for the critical density to be reached.

To measure a possible reduction in the rate of recession means looking back over billions of years. Astronomers must compare the velocities of nearby and distant galaxies. At the same time, in estimating distances, allowance must be made for the more distant galaxies being brighter than nearby ones, in absolute terms, simply because they are being seen much earlier in the history of the Universe and it is assumed that, owing to the evolution of the stars in the galaxy, its brightness will diminish with time.

Of course, at a very early stage as stars are born from the condensing gases, a galaxy would brighten with time, and its average brightness may remain fairly constant for millions of years before stars decay in sufficient numbers to affect the overall brightness of the galaxy. The limits of error are well outside those needed to prove whether the Universe is open or closed, from all the data available so far, but, on average, the observed velocities seem to exceed those required for escape. We can, however, set the extreme limits to Hubble Time, and to the rates of deceleration of the galaxies. What is needed is some form of constraint on the amount of matter that may be present in intergalactic space.

At the instant of the big bang, the entire Universe consisted of neutrons, protons and electrons, it is supposed. At the temperatures and densities present in the first few minutes, the protons and neutrons would have fused to form deuterium nuclei. Deuterium is an isotope of hydrogen, known best perhaps for its presence in 'heavy' water (D_2O). Within a short while, the conditions for nuclear fusion would have ceased, and the expanding fireball would continue to produce hydrogen and helium, and some of the deuterium nuclei would be fused into helium nuclei. The proportion of helium and deuterium remaining gives an indication of the initial density of the Universe, and from the newly-discovered microwave background temperature we can deduce what the present density should be.

Observations of deuterium density in interstellar space from satellite observatories in recent years have indicated that the present density of the deuterium is 4×10^{-28}kg/m^3. Although this may only be an approximate measurement, theory shows that, if the Universe had been ten times denser (for example), the deuterium would have been a thousand times rarer, clearly suggesting that the density assumed is about right. Of course, it does depend critically on the value of Hubble Time, but this would have to be considerably different to discredit the evidence from deuterium measurements that the Universe is open and infinite.

However, the theoreticians have not finished by a long chalk! In 1977, two physicists, Benjamin Lee and Steven Weinberg suggested that there was evidence for the existence of 'heavy neutrinos', atomic particles that could exist in sufficient quantities between the galaxies to cause the mass of the Universe to exceed the critical value.

The answer will probably not be known with any certainty for a long time. All that can be said at present is that the origin of the Universe, and its present limits, can only be guessed at with a slight chance that we are more likely to be somewhere near the correct answer than not!

5 Is Anybody There?

The question of whether there is life anywhere in the Universe besides Earth is not, strictly speaking, an astronomical one at all. But because many biologists' philosophies are rather hostile to the possibility of life elsewhere in the Universe, it helps, perhaps, if the astronomer tries to redress the balance.

As it is not strictly astronomy, the astronomer may ask: who cares if there is anyone 'out there'? But it can scarcely be denied that evidence of life on other worlds would have a profound effect on the human race—more profound still if there was evidence of intelligence elsewhere in the Universe. Sadly, one may have to answer the question 'Is there intelligent life on Earth?' before one can speculate on whether such knowledge would have a good or bad effect. But the questing mind of Man will not stop looking for answers to these questions: and it is this which makes astronomy such a fascinating subject.

As yet, there is no evidence on Earth for the existence of alien beings. Before the flying-saucer enthusiasts raise howls of protest, it must be clearly stated that UFOs (flying saucers) have been seen many times, and many of the sightings cannot be explained, but that does not make them space ships from other worlds. Readers of this book may ask 'Which world?' because, from the perspective of the Universe that should have been perceived from these pages, the answer is clearly not a world in this Solar System, neither is it a world of a binary star, nor a variable star, nor a star of radically different stellar class to our own Sun, which rules out virtually all the stars within 4pc of the Sun. Once we are talking about distances greater than this (or even as great as this) the logic of how journeys lasting hundreds of years can be worth even the energy expenditure alone, simply to flit about the upper atmosphere of the Earth and make no serious contact with the species below defies any serious explanation. Flying saucers have been seen, but they are not what some people would like to think they are.

They are natural, sometimes very peculiar, but none the less ordinary phenomena, ranging from inversion layers in the atmosphere, artificial satellites, meteorological balloons and aircraft, to the planet Venus!

The only example of the evolution of life that we have is on the Earth itself. The exploration of the other planets in recent years has painted an increasingly gloomy picture of the possibility of life evolving in conditions much different from those on Earth, although, at the time of writing, the final verdict on Mars has not been delivered. The biologists may say 'I told you so.' Life, they argue, has evolved through a number of coincidences. The right chemicals were present in the primal atmosphere of the Earth, with the right temperature and chemical environment necessary to produce the first organic compounds, and subsequently the right conditions for these to develop to the point at which they could be said to be living, and for the right conditions for this evolution to continue, and for the many thousands of dead ends evolution went through to produce the more complex lifeforms, particularly the most complex of all, those displaying intelligence. If you multiply all the odds of each of these coincidences taking place together, the evolution of Man has taken place against odds of many millions to one.

This, of course, is just one interpretation of the story. It seems to suggest that life is reluctant to emerge, being put off by the slightest hostility in the environment. On the other hand, it can be argued that, given something like the right conditions, and millions of years to experiment, life *will* emerge. Whether it thrives is another matter. But the odds in favour of finding suitable conditions somewhere in our Galaxy (let alone the others) are even greater than the most pessimistic odds offered by the biologists.

The astronomical evidence is slim, but it is growing. In Chapter 3 we saw how molecules of relatively complex compounds (compared with atoms of ordinary elements) have been found in areas such as the emission nebulae, or HII regions. The number of such compounds known before radio astronomy began in earnest was small. Radio astronomers discovered the hydroxyl radical (OH) in the HII areas usually, and surprisingly, as an emission feature of the spectrum of the region. This was due to the maser effect described on page 131, which may be associated with the birth of new stars.

Since then, radio astronomers have identified molecules of ammonia (NH_3) formaldehyde (H_2CO), water, hydrogen cyanide (HCN) and many others (over 40), and the majority of these are organic compounds. Most of the compounds have been found in the regions of greatest concentrations of dust and gas, as might be expected. One HII region near the centre of the galaxy, known as Sagittarius B2, is a cloud some 20 light years across with a density of up to ten million particles per cubic centimetre, at least a hundred times denser than any other known cloud between the stars. More different molecules have been observed in this cloud than anywhere else in space.

These are the compounds that would have abounded in the solar nebula, the cloud of matter which collapsed under gravity to form the Sun and the bits left over, the planets. The planets, therefore, would have acquired many of these compounds in their atmosphere from space as well as from their interiors.

A. I. Oparin and J. B. S. Haldane proposed that the primitive atmosphere of the Earth, subjected to the incident energy that was pouring in from the Sun and being trapped by the dense atmosphere would result in tremendous electrical storms. The conditions would result in the production of many complex organic molecules.

Laboratory experiments in which mixtures of gases and compounds resembling the composition of this primitive atmosphere have been subjected to ultraviolet radiation and electrical discharges have produced amino acids—the compounds found in nucleic acids, which are the basis of living cells.

Throughout its history, the Earth has been bombarded with matter from the depths of space, ranging from molecules of gas to tonnes of rock and iron in meteorites. Meteorites, as we have seen, may result from the break-up of comets, and the comets in their turn may be formed of the primitive stuff of the original solar nebula, or may even originate in interstellar space. If so, the material in many meteorites may not originate in this Solar System at all.

Life may also, it has been suggested, have originated in the interstellar medium and been carried to Earth, and of course planets throughout the Galaxy, by gas clouds, comets, or any form of debris.

Chandra Wickramasinghe, Professor of Applied Mathematics and Astronomy at University College, Cardiff, and Sir Fred Hoyle (the latter being the astronomer at the centre of the controversy about the continuous-creation theory of the Universe) put forward a startling suggestion that the compounds found in the HII regions might stick together during the formation of stars in the collapsing clouds, and as they collided with other clumps of molecules, would form grains some 10 micrometres in size. Many would be destroyed by the star around which they were forming, but many others would be swept away by the growing stellar wind from the central body. Such grains have been tentatively identified in meteorites, and ultraviolet spectra have shown absorption bands that correspond closely with laboratory spectra for organic compounds and with the spectra of material from certain meteorites.

Wickramasinghe and Hoyle therefore suggest that life begins in the space between the stars, in regions like the Orion nebula. Organic molecules of increasing complexity are formed in these clouds, sticking to inorganic dust particles where they are polymerised to a sticky tar-like coating. These clump together with other sticky particles, and react with them, so that life begins its evolution in the very depths of space.

Although this theory is highly speculative at the moment, there are observational data which support it. Some meteorites have shown evidence of fossil cell-like structures embedded in the inorganic material which resemble the microfossils found on Earth.

The theory was extrapolated further by these two astronomers with a suggestion that interstellar matter could still be carrying alien bacteria or viruses that, on the rare occasions that such bodies enter the Earth's atmosphere, could cause epidemics in various plants and animals. Such an origin of new bacteria would explain why epidemics break out without any particular reason simultaneously in different regions on Earth at apparently random intervals of history.

Hoyle and Wickramasinghe also challenge the Oparin and Haldane theory of the evolution of life in the Earth's primal atmosphere. Astronomical evidence, they say, suggests an early atmosphere with too low a proportion of hydrogen to prevent oxidation of the organic compounds, which would render the suggested evolution impossible.

Another contribution to this highly contentious debate has been made by Clifford Matthews and Robert Minard of the Illinois and Pennsylvania State Universities, and their colleagues. The energy required to form chains of protein from amino acids is very high, so there is no obvious way that the crude compounds formed by discharges in the pre-biotic atmosphere could have formed proteins, as they would presumably have been deposited by rain into the oceans that covered most of the planet. But Matthews and Minard discovered a low-energy reaction involving hydrogen cyanide and water that produces suitable compounds for the formation of proteins, and which would probably have been the origins of the amino acids in the earlier pre-biotic-atmosphere simulation experiments. Such a reaction would also explain the amino acids found in meteorites, since hydrogen cyanide is plentiful in space. (The Orion nebula has a number of HCN clouds in its central regions.)

The important feature of many of the most promising explanations of the origins of life is that the conditions will arise in many thousands of millions of solar systems. But advanced forms of life and intelligent creatures would need rather more comfortable conditions than the average bacterium, and, as we have seen, that rules out many types of star.

Life on Earth has evolved through a long history of trial and error, and although many species flourished by adapting to special conditions, such as mountain sides, or extremely cold regions, or the darkest depths of the oceans, there were plenty of places on Earth where the environment defied the evolution of all but the simplest creatures. So, although some strains of bacteria can survive boiling nitric acid, in general life based on the chemical building blocks we have discussed seems to be confined between certain extremes of temperature, pressure and (perhaps) gravity. But even with such constraints, many scientists now believe that life must be common throughout the Galaxy and that there must be many examples of advanced civilisations.

It is only playing with figures, but to put the chances in perspective assume that only one star in ten has planets, only one in ten of these has planets that could sustain life, only one in ten of these actually does sustain life, only one in ten of these has advanced life, only one in ten of these has advanced life with intelligence, only one in ten of these has developed advanced civilisations, and only one in ten civilisations are contemporary with ours, then there are ten thousand civilisations at present in the galaxy.

Still playing with figures, if these civilisations are fairly evenly distributed about the galaxy, the nearest civilisation to our own would be, on average, about 250 parsecs away. This means, of course, that to send them a message and receive their reply would involve a wait of 1630 years, so if the ancient Romans had sent out Man's first message to the stars, 'How are you?', in 350AD, any moment now our radio telescopes would pick up the alien equivalent of 'Fine! How're you?' any minute now.

Of course, many people believe that we should not announce our presence to the Universe, until we know what our neighbours' intentions might be. But Man has already done that. Marconi's transatlantic signals have by now reached some 24 parsecs into space. The frustrations of trying to make such an announcement deliberately, when the reply might not come back for fifty generations or more, suggest that it is of little hope to search for signals from some alien civilisation deliberately transmitted to call the attention of the outside worlds. The best we could hope for is to pick up the equivalent of our commercial radio broadcasts.

Unfortunately, the likelihood of a civilisation being in our neighbourhood is almost impossible to estimate. While some factors in the odds quoted earlier may be too pessimistic (for example, once life emerges, perhaps intelligence and civilisations would follow in most cases) other factors will offset these (the existence of intelligent life has been but a split second in the Earth's lifetime—what chance is there that another star, that could have formed millions of years before or after our Sun, would also have inhabited planets?).

Many radio telescopes are now devoting some time to the fascinating possibility of picking up signals that are not of natural origin. Project SETI (search for extraterrestrial intelligence) has been considered by NASA, based on the use of a very large radio receiving array searching those wavelengths that a civilisation sending out a 'Here we are!' signal would probably choose. An alternative strategy has been proposed at the Pasedena Jet Propulsion Laboratory in which the whole sky and spectrum is searched for any anomalous radio signals.

The final question is the most difficult. When Man first picks up signals from another civilisation, he will know that his planet, his Sun, and his species are in no way the centre of the Universe; can he survive that knowledge?

Postscript: Down to Earth

Astronomy has made advances in recent years which exceed any made in its entire history. But for all that, who is to say that we are significantly nearer the truth than was Ptolemy?

Much of our thinking is based on flimsy models, even flimsier evidence, and a great deal of guesswork. But, if it has taught us nothing else, modern astronomy at least has given us a sense of perspective, and a flexibility of approach.

While black and white holes are, as yet, examples of astronomers' imagination, and white holes in particular are very speculative suggestions, they are worthy of serious consideration. There is every good reason for a scientist to pursue his own special hobby in astronomy, like proving the Earth is flat, the Universe closed, the sun-spot cycle a recent phenomenon, that the Moon craters were mostly volcanic or impact, and so on. There is so much to find out, that such special questions must be left to individuals until such time as they can convince their fellow astronomers.

There is much for other specialists to contribute. Increasingly, the astronomer needs to consult the geologist, the biologist, and, one day perhaps, even the linguist. The ordinary amateur can also provide valuable information, once he has an idea what may be of interest. It is hoped that this book may have helped him to understand what the problems are, and to try to help others perceive some of the wonder of the Universe. In our supposedly enlightened era, parascience flourishes. The sensation-hungry public will more readily soak up rubbish such as flying-saucer visitations, alien visitors in the Earth's pre-history, astrology, Bermuda Triangles and the like than make the effort to think about the *real* mysteries of the Universe, which are sufficient for anyone's intellectual exercise, or just simple enjoyment, for a lifetime.

Appendix

Making and using a simple telescope

Much of what has been described in this book can be observed with a small telescope, or even a good pair of reasonably high-powered binoculars. The use of a tripod is essential for all instruments being used for astronomy as it is impossible to see anything like the full detail available when the field of view is dancing about, as is inevitable with hand-held binoculars or telescope.

It is quite easy to devise a clamp to fit around the centre focusing bar of a pair of binoculars which can be used to mount them on a sturdy photographic tripod—or even a home-made stand.

To make a telescope means investing in a reasonable quality objective lens, although cheap lenses can be used if nothing else is available, and the resulting instrument quite capable of showing the easiest objects, such as Jupiter's satellites, Moon craters, etc. If possible the lens should be over 60-mm diameter, because you will probably have to mask off the outer edge to reduce aberration, and simply to hold the lens in place.

Very successful telescopes can be made with plastic piping, of the type sold for rain water pipes, waste pipes and so on. These are easy to joint, cut, to make sliding fits one tube inside another and are cheap for a material which is rigid and does not corrode.

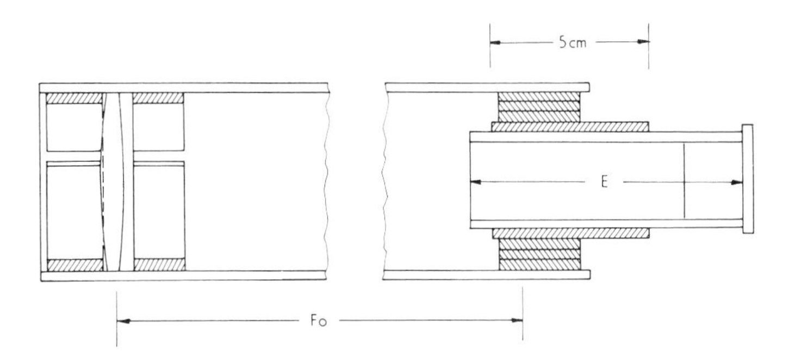

34 Construction of a simple refractor from plastic piping of different diameters. The eyepiece tube should be of length E equal to the focal length plus about 2cm. The sliding-fit tube in which the eyepiece slides must be held in suitable packing in the main tube, the packing being centred at a distance from the objective equal to its focal length F_0.

The interior of the tubes should be blackened with matt-finish paint before mounting the lens. This is usually possible by fixing near the end of the tube (it should be mounted about 60mm inside the tube, or a hood of at least this long added afterwards) with two rings cut from another tube of different diameter which if possible is a sliding fit inside the main tube. If the inside tube is too small to make a sliding fit, it can be cut axially, opened out and, if necessary, glued in position. With a number of rings of this sort the inner diameter of the telescope tube can be reduced to make a snug fit on the objective, and additional rings bolted either side of the lens to secure it in position.

The eyepiece should be similarly secured inside a separate tube which is a sliding fit in the objective tube. This is worth getting just right, rather than using packing pieces inside the larger diameter tube, since this sliding fit will be used to adjust the relative positions of the objective and eyepiece to find the focus.

The extra length of tube at the objective end protects the objective from stray light, and from dewing when humidity is high. It is helpful to let telescopes cool down during the evening before use, and even more important to protect the objective when it is cold after a night's use and it is brought into a warm room.

The magnification of a telescope is the ratio of the objective focal length to the eyepiece focal length, but it is more important to gather light, and to be able to see the image clearly, which requires as large and as high-quality an objective as money will allow.

Very high-power eyepieces are usually difficult to use, lose much light, and increase the problems of steadying the telescope. An objective of about 500mm focal length and an eyepiece of about 25mm will give a magnification of 20 times and a good bright field covering a reasonable angle of sky. Such a telescope will show the rings of Saturn, and separate many double stars.

If you are spending a lot of money on optics, it is worth going into the construction of amateur telescopes in some detail, and some of the books in the bibliography will help here. Larger amateur telescopes are usually reflecting telescopes with a special mirror as the objective, rather than a lens as in the refractor. Reflectors are even easier to make than refractors at any size over about 100mm aperture. But again, it is worth aiming for a fast objective, say about f5 or f6, than making or buying long-focus objectives that might seem to give high magnification, but which are very difficult to use, give a dim image, and are generally frustrating. The focal ratio (f number) of a lens is familiar to photographers, and indicates the speed of the lens: the lower, the faster. It is the ratio of the focal length to the diameter of the lens or mirror. Thus a 200-mm diameter mirror with a focal length of 1.2m has a focal ratio of f6.

Using a telescope

There is quite an art in using a telescope, particularly in finding the object you are looking for from the position given in an atlas. It is well worth training yourself to know the daily and monthly changes in the position of objects in the sky and to learn systematic methods of finding objects in the telescope.

Further Reading

The following selection of publications is intended to give the reader only a guide to the many possible avenues of interest that he or she may now wish to follow.

For news and up-to-date reports on discoveries in astronomy, the author can unhesitatingly recommend the reader not to believe what is published in the daily press, although such items are useful as tips to read the facts in more specialist periodical publications. These include:

Sky and Telescope, Sky Publishing Corp. 49–51 Bay State Rd., Cambridge, Mass. (Monthly: the essential publication for combining practical and theoretical work.)

Nature, 4 Little Essex Street, London WC2R 3LF.

Science, The American Association for the Advancement of Science, 1515 Massachusetts Ave., NW, Washington D.C. 20005.
(Both the above publications are weekly and give material direct from the observatories written by professional astronomers.)

Scientific American, 415 Madison Avenue, New York, N.Y. 10017. (Monthly: very high quality features on physics and astronomy as well as all other sciences.)

New Scientist, King's Reach Tower, Stamford Street, London SE1 9LS. (Weekly: very rapid news coverage, and useful features.)

Among the semi-professional and learned-society journals, and similar types of publications, the following should be mentioned:

The Astrophysical Journal (and *Journal* Supplements), University of Chicago Press, 11030 Langley Ave., Chicago, Illinois 60628.

Journal of the Royal Astronomical Society, Burlington House, London W1V 0NL.

Journal of the British Astronomical Association, Burlington House, London W1V 0NL.

Journal of the Royal Astronomical Society of Canada, 252 College Street, Toronto 2B, Ontario.

Some of the selection of books which follows will suit newcomers to the subject who wish to expand their knowledge. Others are intended for readers who wish to advance their existing knowledge:

A useful review of the history of astronomy is given in *The Astronomers*, Colin Ronan, (Evans Bros., London) and in other books by the same author. The classic book on the history of astronomy up to Newton's time is *The Sleepwalkers* by Arthur Koestler.

Much of the astronomy of the Solar System is covered in the more general books listed later, but the best up-to-date work is *The Solar System* (originally published in *Scientific American* September 1975) (W. H. Freeman & Co., San Francisco and Reading.)

Moving out among the stars and galaxies, the pulsars are well covered in *Pulsars*, F. Graham Smith (Cambridge University Press, London), and this should be followed by *Black Holes in Space*, Patrick Moore and Iain Nicolson (Orbach and Chambers, London). More general coverage is given by *Exploring the Galaxies*, Simon Mitton (Faber and Faber, London), which goes into the subject in some depth, and for the most detailed (and fairly expensive) material by leading authorities see *Structure and Evolution of Galaxies*, edited by G. Setti (D. Reidal). *The Radio Universe*, 2nd ed., J. S. Hey (Pergamon Press, Oxford) is excellent for the reader with only average scientific background.

On cosmology, *Astronomy and Cosmology—a modern course*, Fred Hoyle (W. H. Freeman & Co., San Francisco and Reading), is a standard work, and covers the whole subject as well as cosmological aspects. Also see *Modern Cosmology*, D. W. Sciama (Cambridge University Press, London).

In general vein, the following are very good value:

Our Changing Universe, John Gribbin (Macmillan, London); *Planets, Stars and Galaxies*, 4th ed., Stuart J. Inglis (John Wiley & Sons Inc., New York and London); *Vistas in Astronomy*, by Arthur and Peter Beer, (eds.) (Pergamon Press, Oxford, published periodically); *The Cambridge Encyclopaedia of Astronomy*, Simon Mitton, (ed.) (Jonathan Cape, London); *Astronomy Today*, Fred Hoyle (Heinemann Educational Books, London).

A popular review of the search for, and the prospects of finding, life in the universe, is given in *Worlds Beyond*, Ian Ridpath (Wildwood House, London).

For those who wish to follow up the practical aspects of astronomy, and the use of telescopes, the author is familiar with *Discover the Sky with Telescope and Camera*, Richard Knox (R. Morgan, Chislehurst, Kent), and *Experiments in Astronomy for Amateurs*,

Richard Knox (David and Charles, Newton Abbot, Devon), and can also recommend *Practical Amateur Astronomy*, Patrick Moore (ed.) (Lutterworth, London). For those who need positional and other material from an almanac, the *Handbook* of the British Astronomical Association (see above) is outstanding value and the standard works are: *The Astronomical Ephemeris* and the *American Ephemeris and Nautical Almanac* and the *Explanatory Supplement* (HMSO, London and US Govt. Printing Office, Washington).

Index